High-Speed Devices

Devices

and Circuits

with THz

Applications

Devices, Circuits, and Systems

Series Editor
Krzysztof Iniewski
CMOS Emerging Technologies Research Inc.,
Vancouver, British Columbia, Canada

PUBLISHED TITLES:

FORTHCOMING TITLES:

3D Circuit and System Design: Multicore Architecture, Thermal Management, and Reliability
Rohit Sharma and Krzysztof Iniewski

Circuits and Systems for Security and Privacy
Farhana Sheikh and Leonel Sousa

CMOS: Front-End Electronics for Radiation Sensors
Angelo Rivetti

Gallium Nitride (GaN): Physics, Devices, and Technology
Farid Medjdoub and Krzysztof Iniewski

High Frequency Communication and Sensing: Traveling-Wave Techniques
Ahmet Tekin and Ahmed Emira

Labs-on-Chip: Physics, Design and Technology
Eugenio Iannone

Laser-Based Optical Detection of Explosives
Paul M. Pellegrino, Ellen L. Holthoff, and Mikella E. Farrell

Metallic Spintronic Devices
Xiaobin Wang

Mobile Point-of-Care Monitors and Diagnostic Device Design
Walter Karlen and Krzysztof Iniewski

Nanoelectronics: Devices, Circuits, and Systems
Nikos Konofaos

Nanomaterials: A Guide to Fabrication and Applications
Gordon Harling, Krzysztof Iniewski, and Sivashankar Krishnamoorthy

Optical Fiber Sensors and Applications
Ginu Rajan and Krzysztof Iniewski

Organic Solar Cells: Materials, Devices, Interfaces, and Modeling
Qiquan Qiao and Krzysztof Iniewski

Power Management Integrated Circuits and Technologies
Mona M. Hella and Patrick Mercier

Radio Frequency Integrated Circuit Design
Sebastian Magierowski

Semiconductor Device Technology: Silicon and Materials
Tomasz Brozek and Krzysztof Iniewski

Soft Errors: From Particles to Circuits
Jean-Luc Autran and Daniela Munteanu

High-Speed Devices

and Circuits

with THz

Applications

Edited by **Jung Han Choi**
Fraunhofer Institute, Berlin, Germany

Krzysztof Iniewski Managing Editor
CMOS Emerging Technologies Research Inc.
Vancouver, British Columbia, Canada

CRC Press
Taylor & Francis Group
Boca Raton   London   New York

CRC Press is an imprint of the
Taylor & Francis Group, an **informa** business

CRC Press
Taylor & Francis Group
6000 Broken Sound Parkway NW, Suite 300
Boca Raton, FL 33487-2742

First issued in paperback 2017

Version Date: 20140415

ISBN 13: 978-1-4665-9011-3 (hbk)
ISBN 13: 978-1-138-07158-2 (pbk)

Library of Congress Cataloging-in-Publication Data

High-speed devices and circuits with THz applications / editor, Jung Han Choi.
   pages cm -- (Devices, circuits, and systems ; 30)
   Includes bibliographical references and index.
   ISBN 978-1-4665-9011-3 (hardback)
   1. Terahertz technology. 2. Electronic circuits. 3. Very high speed integrated circuits.
I. Choi, Jung Han.

TK7877.H54 2014
621.381'3--dc23   2014003132

Visit the Taylor & Francis Web site at
http://www.taylorandfrancis.com

and the CRC Press Web site at
http://www.crcpress.com

Contents

Preface

This book was motivated by the desire to explore the future perspectives of high-frequency technologies by presenting the state-of-the-art results in both new device developments and circuit implementations. It discusses issues for the circuit operation beyond 100 GHz with respect to device physics, circuit implementations, and some technological bottlenecks for system implementations. Also, recent device and circuit results using SiGe technologies as an alternative to Si complementary metal-oxide-semiconductor (CMOS) devices are presented.

Evolutionary device developments let people consider widening the frequency spectrum of interest up to terahertz. For example, as Si CMOS technologies evolve from 65 nm to 40 nm to 28 nm and further, the expected operation frequency of circuits correspondingly increases beyond 100 GHz. Nowadays, several universities and companies pursue developing high-frequency circuits and subsystems that could operate 60, 77, 90, 120, and 300 GHz.

Further challenging research works on nano-electronic devices have been reported, even enabling terahertz operations. Several research directions are challenged. New nano device developments using nanowires or carbon-based elements, e.g., graphene field effect transistors (FETs), carbon nanotubes, etc., seem to be very active research themes around the world in terms of new radio frequency (RF) directions. As the device technology continues to develop, circuit engineers could design integrated circuits (ICs) with higher operation frequencies and develop relevant new applications. It seems that novel devices could lead to the creation of new applications. Also, market needs would guide the direction of future developments, e.g., lower power consumption, higher integration density, and higher speed of operations.

Another exciting technical trend that we observe in the market and experience in daily life is the so-called big bang of Internet data traffic caused by handheld devices, such as laptops, tablets, and smart phones. As the amount of data traffic increases, higher-speed data transmitters and receivers play a greater role in the higher data transfer communications and now get meaningful attention from RF industries. For example, connectivity technology between high-performance computing servers in the data center becomes an important issue with respect to data transfer rates, distance, power consumption, the cost of connectivity modules, and backward/forward compatibilities with concurrent standards technologies. Several companies and research institutions exert their efforts to demonstrate next-generation connectivity technologies to customers.

This book tries to put those modern topics together, e.g., novel/new nano devices, their feasible THz applications, and modern high data transfer technologies. By bringing those heterogeneous topics into one, engineers who work in high-frequency engineering fields or develop devices and circuits considering high-frequency applications are supposed to grasp modern technical trends of nano devices, circuits, and their applications. The contributions are made by several academic institutions and relevant companies. Readers will grasp the future potential of RF products and research trends by reading through several chapters by the international experts in industry and academia.

Chapter 1 is contributed by a researcher at the Tokyo Institute of Technology and discusses THz sensing and imaging devices based on nano devices and materials. Various devices are explained in detail, especially regarding THz imaging. Chapter 2 is written by a researcher at Université Catholique de Louvain, which investigates SOI multigate nanowire FETs. It explains theoretical aspects of nanoscale nanowire MOSFETs, simulation methods, and their results. Chapters 3 and 4 are then devoted to SiGe technologies. They thoroughly discuss the physics of the SiGe heterojunction bipolar transistor (HBT) and present commercially available SiGe HBT devices. Also, several feasible applications using those devices are addressed. Very recent experimental results over 100 GHz are presented. If readers are interested in designing ICs working at beyond 100 GHz, it is recommended to read through those two chapters written by SiGe device experts. Chapter 5 is also very instructive in terms of THz IC design using standard Si CMOS devices, especially for THz imaging application. The author discusses in detail experimental setups for measurements, detection methods, and so on. Chapters 6 to 9 are devoted to high-speed data rate connectivity technologies. Three chapters are contributed by industrial colleagues and one by the University of Toronto. The discussion points range from system design to IC design. Also, they are useful to understand current state-of-the-art technologies in these development fields. Also, relevant standard activities and technical details behind those are addressed. One can have an outlook over these interconnection technology trends by reading those chapters.

It is impossible for me to express my gratitude adequately to all contributors for their sincere and passionate preparation of their chapters. I will remember their contributions and active exchange of their thoughts and ideas. Also, thanks should be given to Dr. Kris Iniewski for giving me an opportunity to edit this comprehensive and valuable book.

Finally, I also thank my lovely wife, Mrs. Hye Won Nam, for her warm support in my completing this book.

Dr. Jung Han Choi
Charlottenburg, Berlin, Germany

About the Editor

Jung Han Choi received B.S. and M.S. degrees in electrical engineering from the Sogang University, Seoul, Korea, in 1999 and 2001, respectively, and the Dr.-Ing. degree from the Technische Universität München, Munich, Germany, in 2004. From 2001 to 2004, he was a research scientist in the Institute for High-Frequency Engineering at the Technische Universität München. During this time, he worked on high-speed device modeling, thin-film fabrication, network analyzer measurement, and circuit development for high-speed optical communications. From 2005 to 2011, he was with the Samsung Advanced Institute of Technology and the Samsung Digital Media & Communication Research Center, where he worked on the radio frequency (RF) biohealth sensor, nano devices, and RF/millimeter-wave circuit design, including 60 GHz Si complementary metal-oxide-semiconductor (CMOS) integrated circuits (ICs). In 2011 he joined the Fraunhofer Institute (Heinrich-Hertz Institute), Berlin, Germany. Now he is working on high-data-bit-rate transmitter and receiver circuits up to 100 Gb/s, relevant device active/passive modeling, and network analyzer measurement up to 170 GHz.

In 2003, he was awarded the EEEfCOM (Electrical and Electronic Engineering for Communication) Innovation prize for his contribution to the development of the high-speed receiver circuit. His current research interests range from the active/passive device, carbon-based nano device, and its modeling to high-frequency IC design and metamaterials.

Contributors

Aryan Afzalian
ICTEAM Institute
Université Catholique de Louvain
Louvain-La-Neuve, Belgium

Anthony Chan Carusone
Department of Electrical and Computer
    Engineering
University of Toronto
Toronto, Ontario, Canada

Yongmao Frank Chang
Vitesse Semiconductor
Camarillo, California

Jung Han Choi
Fraunhofer Institute
Heinrich-Hertz Institute
Photonic Components
Berlin, Germany

Péter Földesy
Computer and Automation Research
    Institute
Hungarian Academy of Sciences
Budapest, Hungary
and
Faculty of Information Technology
Péter Pázmány Catholic University
Budapest, Hungary

Yukio Kawano
Department of Physical Electronics
Quantum Nanoelectronics Research
    Center
Tokyo Institute of Technology
Tokyo, Japan

Rohit Mittal
Intel Corporation
Santa Clara, California

Edward Preisler
Jazz Tower
Newport Beach, California

Marco Racanelli
Jazz Tower
Newport Beach, California

Jae-Sung Rieh
School of Electrical Engineering
Korea University
Seoul, Korea

Jafar Savoj
Xilinx
San Jose, California

1 Terahertz Technology Based on Nano-Electronic Devices

Yukio Kawano

CONTENTS

1.1 INTRODUCTION

Since the establishment of the Maxwell equation and the discovery of the electromagnetic wave, researchers have been devoted to developing a lot of technologies based on the electromagnetic wave, bringing much change to human life. High-frequency electronics technology has provided radio, television, the cellular phone, and so on.

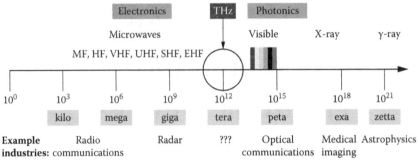

FIGURE 1.1   Chart of the electromagnetic wave spectrum. (Adapted from the THz Science & Technology Network [http://thznetwork.net/].)

From optics and photonics, optical communication, light-emitting diode (LED) lamp, endoscope, etc., have been produced.

The terahertz (THz, 10^{12} Hz) frequency region is located in between the microwave region and the visible light region (Figure 1.1). This region was merely studied by a small number of researchers in limited fields, such as chemical spectroscopy, astronomy, and solid-state physics. However, THz technology is nowadays in strong demand in a large variety of fields, ranging from basic science such as biochemical spectroscopy, astronomy, and materials science to practical science such as environmental science, medicine, agriculture, and security [1, 2]. What is the reason why the THz wave has attracted so much interest? The advantageous properties of the THz wave are that it can be transmitted through objects opaque to visible light and that the corresponding photon energy, 1–100 meV, is in the important energy spectrum for various materials and biomolecules. These features allow various applications of imaging and spectroscopy in this frequency band. Figure 1.2 displays several examples of applications of the THz waves. The measurements shown here illuminate the THz waves onto objects and map intensity distributions of reflected or transmitted THz waves. In some situations, by simultaneously measuring frequency spectra, one is able to identify the contents and characterize their physical/chemical properties. The technique therefore can be used as nondestructive inspection, which is based on the fact that the THz wave is much safer and does not do much damage, compared to the x-ray. In addition to industrial and medical applications, THz technology is also of much importance in basic sciences. For example, in astronomy, materials science, and biochemistry, the detection of very weak THz radiation from interstellar matter in space, electrons in materials, and biomolecules is expected to unlock mysteries behind the generation of the universe, quantum effect in materials, and life activity, respectively.

In the THz region, however, even basic components like detector and source have not been fully established, compared to the technically mature, other frequency regions. This is because the frequency of the THz wave is too high to be handled with conventional high-frequency semiconductor technology. In addition, the photon energy of the THz wave is much lower than band gap energy of semiconductors.

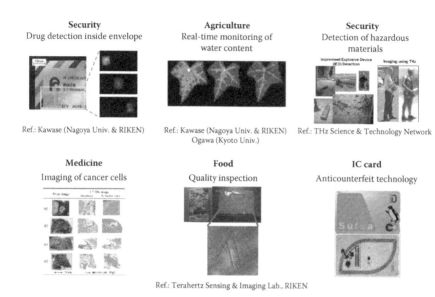

FIGURE 1.2   Various applications of THz imaging and spectroscopy.

For the above reasons, the THz wave is not easy to approach from either side of electronics and optics/photonics. In imaging technology, the THz wave also has a problem of low spatial resolution, which results from much longer wavelengths of the THz waves compared to those of visible light.

The applications of nanoscale materials and devices, however, are opening up new opportunities to overcome such difficulties. Nanostructured devices based on the superconductor, semiconductor, and carbon nanotube have enabled significant improvement in detection sensitivity and spatial resolution. In this chapter, I will describe new THz sensing and imaging technology based on such nano-electronic devices. Moreover, I will show applications of cutting-edge THz measurements to materials researches, which have provided new insight into electronic properties of the materials.

1.2   THz DETECTOR

1.2.1   Overview

THz detectors can be generally categorized into three types: bolometric (thermal) detection, wave detection, and quantum detection. In addition, they are also used as photoconductive antenna, electro-optic device, frequency mixer for heterodyne detection, etc. In this section, I show an overview of various THz detectors and briefly discuss their advantages and disadvantages.

1.2.1.1   Bolometric (Thermal) Detection

This type of detector utilizes temperature rise via THz absorption. As the crystal temperature of the detector decreases, the detection sensitivity is improved, but the detection speed becomes low (a typical time constant is of the order of ms below 4 K). The detected signal is mostly resistance change arising from the rise in the crystal temperature due to the THz absorption. There is another type of readout mechanism: measuring gas pressure via thermal expansion (Golay cell detector). Since all the detectors based on bolometric detection respond to electromagnetic waves in a wide-frequency region, one needs a frequency cutoff filter. Wei et al. [3] recently reported a nanoscaled superconductor bolometer with the ability of a single THz photon detection.

1.2.1.2   Wave Detection

This detector senses the THz electromagnetic wave as a high-frequency wave. A representative device is a Schottky barrier diode detector. The advantages of this detector are high-speed detection (time constant of ~ns) and room temperature operation. This detector is often used in sub-THz regions, because the sensitivity becomes low with increasing the frequency of the incident THz wave.

1.2.1.3   Quantum Detection

In contrast to wave detection, this type of detector senses photons of the THz electromagnetic wave. Solid-state devices based on materials like superconductors and semiconductors usually have energy level spacing corresponding to the THz photon energy (1–100 meV), for example, energy gap of superconductor, impurity level of semiconductor, and energy level spacing due to quantum electron confinement of semiconductor quantum structure. It follows that excess carriers are generated in the devices, when these devices absorb the THz waves. One can detect the THz wave by recording electric signals produced by the carriers. THz detectors based on a nanoscale island connected to electrodes, such as superconductor junctions [4] and semiconductor quantum dots [5, 6], have been presented and demonstrated to exhibit ultra-high sensitivity, including the level of single-photon detection. However, the operation of the superconductor and semiconductor detectors needs a very low temperature environment (<0.3 K). This situation forces one to use a dilution refrigerator or a ^{3}He refrigerator, restricting the range of practical uses.

1.2.1.4   Others

THz time-domain spectroscopy (TDS) is nowadays widely employed in THz research fields. As detectors, the photoconductive antenna and the electro-optic device are used. This measurement allows real-time observation of the oscillatory electric field of the THz wave, providing information on both amplitude and phase of the THz electric field. Though these detectors work under room temperature, their operation requires an expensive femtosecond pulse laser.

Another type is a frequency mixer using a supercomputer in which a beat signal corresponding to the frequency difference with a local oscillator is measured. This type of detector is often used in the fields of astronomy and environmental science.

1.2.2   CNT-Based THz Detector

The carbon nanotube (CNT) is expected to be used as a building block for future nano-electronics, nano-photonics, and nano-mechanics owing to its unique one-dimensional structure [7]. From the viewpoint of application to the THz detector, the CNT also has been regarded as a promising candidate. In fact, near-infrared sensors using the CNT transistor have been successfully developed [8, 9]. In the THz region, although there are several reports [10, 11], much higher performance has not been achieved to date, compared to other existing detectors.

In the following, I will introduce our recent development in CNT-based THz detectors [12, 13]. We have created a highly sensitive and frequency-tunable THz photon detector using the CNT and GaAs/AlGaAs transistors. This detector has two detection mechanisms: photon-assisted tunneling (PAT) [12] and current-peak shift [13]. For the latter, THz irradiation causes a peak shift of CNT currents relative to gate voltage, leading to the realization of ultra-high sensitivity: THz photon detection.

1.2.2.1 Photon-Assisted Tunneling in CNT

In 1963, Tien and Gordon proposed a theory on the interaction of a nanoscale island with an electromagnetic wave [14]. The essence of their discussion is that new energy bands, the so-called photon sidebands, are formed by the AC electric potential of the electromagnetic wave at intervals of nhf (n, integer number; h, Planck constant; f, frequency of the electromagnetic wave).

A quantum dot (QD) structure connected to source, drain, and gate electrodes is a promising device for examining their model, because the QD with the electrodes works like a single-electron transistor (SET). By sweeping either the gate voltage or the source-drain voltage and measuring the resulting source-drain current, spectroscopic information about energy states in the QD can be directly obtained. The photon sidebands in the QD can thus be observed as new current signals are generated via the inelastic electron tunnel when electrons exchange photons. This phenomenon is known as PAT (Figure 1.3(a)). With microwave irradiation of superconductor tunneling devices [14–16] and semiconductor QDs [17, 18], the PAT has been successfully observed.

Compared to conventional QDs based on superconductors and semiconductors, a QD using a CNT has the following unique properties: its characteristic energy, or charging energy, and energy level spacing due to quantum electron confinement typically reach ~10 meV, corresponding to a THz frequency (~2.4 THz). This energy range is larger by a factor of 10 than that of the conventional QDs. These advantageous properties potentially lead to device applications of CNT-QDs in the THz region. In this view, we have studied the THz response of the CNT-QDs [12].

Figure 1.3(b) schematically displays the CNT-QD structure. The CNT-QD was fabricated as follows: First, CNTs were dispersed on a semiconductor (GaAs/AlGaAs) wafer, whose surface was covered with a SiO_2 film of ~100 nm thickness. We mapped topography of the device surface with an atomic force microscope, and specified locations of single-wall CNTs with a diameter of ~1 nm. Based on the images obtained, source and drain electrodes with an interval of ~600 nm and side-gate electrodes were patterned with electron-beam lithography. In this device, electrons are confined to a very small area of 1×600 nm^2, forming a QD. Metallic CNTs were used.

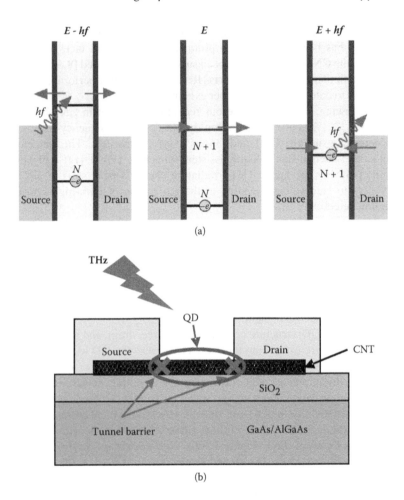

(a)

(b)

FIGURE 1.3 (a) Schematic diagram of electron tunneling processes in a QD under electromagnetic wave irradiation. Middle panel: When the Fermi level in the source lead aligns with a level in the QD, a source-drain current is passed via elastic tunneling. Left and right panels: When electrons interact with photons, a new current flows via inelastic tunneling, i.e., photon-assisted tunneling. (b) Sketch of a carbon nanotube quantum dot (CNT-QD). (Adapted from Y. Kawano et al., *J. Appl. Phys.*, 103, 034307, 2008.)

The CNT device was mounted in a ^{4}He cryostat at a temperature of 1.5 K, and the device was irradiated with a THz wave through an optical window made from a Mylar sheet. As a THz illumination source, a THz gas laser pumped by a CO_2 gas laser was used.

Figure 1.4 shows the data on effect of THz irradiation on the CNT-QD. Differently from the black curve for the data without THz irradiation, new satellite current peaks are generated for the illumination of the THz wave. It is also seen that its position relative to gate voltage shifts in the positive direction with increasing the frequency, f, of the THz wave. The inset of Figure 1.4 displays the energy spacing, $k\Delta V_G$, between

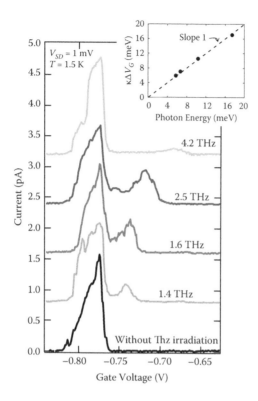

FIGURE 1.4 Source-drain current I_{SD} versus gate voltage V_G for the frequency of the THz wave, $f = 1.4, 1.6, 2.4,$ and 4.2 THz. The experimental curves for the THz irradiation are offset by multiples of 0.8 pA for clarity. The inset shows the energy spacing, $\kappa\Delta V_G$, between the original peaks and the satellite peaks as a function of the photon energy, hf, of the THz wave. The dashed line in the inset is an eye guide corresponding to $\kappa\Delta V_G = hf$. (Adapted from Y. Kawano et al., *J. Appl. Phys.*, 103, 034307, 2008.)

the original peak and the satellite peaks as a function of the photon energy, hf, of the THz wave. From the measurement of the differential conductance dI_{SD}/dV_{SD} as a function of source-drain voltage V_{SD} and gate voltage V_G (Coulomb diamond measurement), we derived the conversion factor, $\kappa = 0.18$, which is defined as the conversion ratio of V_G into energy. The data clearly show good agreement between $\kappa\Delta V_G$ and hf, providing evidence for the THz-PAT in the CNT-QD. It is thus demonstrated that the CNT-QD works as a THz detector with frequency tunability by the application of the gate voltage.

It should be noted that the satellite current is only observed on the positive V_G side with respect to the original peak. The occurrence of the satellite current corresponds to the electron tunneling from the QD to the drain lead (the right side panel of Figure 1.3(a)). The side of V_G on which the satellite current is clearly seen depends on the sample properties. This is possibly due to tunnel barrier asymmetry between the source and drain electrodes.

The theory by Tine and Gordon says that the tunneling rate into the photon sidebands should follow the Bessel functions of the illuminated power [14]. In order to

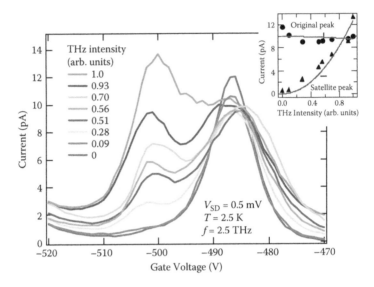

FIGURE 1.5 THz intensity dependence of I_{SD} versus V_G. The inset shows the THz intensity dependence of the amplitudes of a satellite peak and one of an original peak. The solid lines in the inset are fitting curves based on the square of the nth-order Bessel functions. (Adapted from Y. Kawano et al., *J. Appl. Phys.*, 103, 034307, 2008.)

examine this feature, we measured the THz intensity dependence of the THz-PAT in the CNT-QD. In the measurements, we inserted THz attenuating filters between the THz laser and the sample. The result is displayed in Figure 1.5. The general tendency is as expected, but we could not get a perfect fit to the Bessel functions. We speculate that since the single QD was used, the tunneling processes via the electrodes would make it difficult to analyze the result accurately with this model. We plan to study the THz intensity dependence in more detail by using a double-coupled CNT-QD, in which the PAT occurs as a consequence of electron transitions between two well-defined discrete levels. In this device, we anticipate that frequency selectivity would be improved compared to that of the single CNT-QD.

1.2.2.2 THz Photon Detection with CNT/2DEG Hybrid Device

In the CNT THz detector described in the previous section, although frequency-selective THz detection has been achieved, the detection sensitivity is not high. This is because in this detection mechanism, one-photon absorption generates only one electron even if a quantum efficiency of 100% is achieved. This limits the detection sensitivity. In order to resolve this problem, we have created a new designed device [13]: a CNT-SET is integrated with a GaAs/AlGaAs heterostructure chip (Figure 1.6(a)). The hybrid structure is divided into two separate roles of THz absorption (2DEG) and signal readout (CNT-SET). A basic idea of THz detection with this device is that the CNT-SET senses electrical polarization induced by THz-excited electron-hole pairs in the 2DEG. Since the SET works as an ultra-sensitive electrometer, the CNT/2DEG device is expected to exhibit ultimate sensitivity in photon detection.

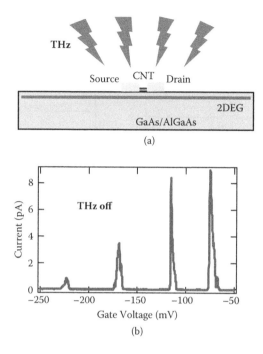

(a)

(b)

FIGURE 1.6 (a) THz detector using the CNT/2DEG hybrid device. (b) Source-drain current I_{SD} as a function of gate voltage V_G of the CNT-SET when the THz irradiation was off.

Figure 1.6(b) shows transport properties of the CNT without THz irradiation; the data of I_{SD} versus V_G at $V_{SD} = 1.5$ meV. This result shows the oscillatory behavior of I_{SD}, indicating that the fabricated CNT device properly functions as a SET.

Figure 1.7 displays the THz response of the 2DEG/CNT device and the dependence on magnetic field, B, applied perpendicular to the 2DEG plane. The experimental setup for the THz measurements shown here is similar to that for the THz-PAT in the previous section. The laser intensity was reduced with a set of THz attenuating filters, and the intensity of the THz radiation applied to the sample is estimated to be 0.75 nW/mm².

The results of Figure 1.7 show that the THz irradiation caused a shift in the current peak position in the direction of positive V_G. The data for the irradiation at 1.6 THz reveal that the shift was monotonically enhanced with increasing B up to 3.95 T, then decreased when B was further raised beyond $B = 3.95$ T. On the other hand, in case of the 2.5 THz irradiation, the B value for the maximum of the peak shift changed to 6.13 T.

The physical meaning behind the features seen in Figure 1.7 is as follows: In the presence of magnetic field, the energy state of the 2DEG forms a Landau level. When the 2DEG is illuminated with a THz wave whose photon energy hf is equal to Landau level separation eB/m^*, the 2DEG absorbs the THz wave (cyclotron resonance) well. Here, e is the elementary charge and m^* is the cyclotron effective mass for the crystal used. The experimental data ($B = 3.95$ T for $f = 1.6$ THz and $B = 6.13$ T for $f = 2.5$ THz) show that the f value is proportional to the B value for the maximum

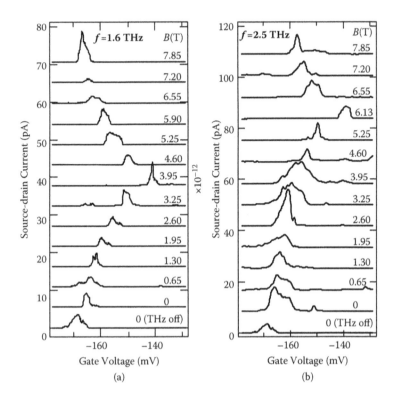

FIGURE 1.7   Magnetic field dependence of source-drain current I_{SD} versus gate voltage V_G of the CNT under THz irradiation. The magnetic field B was applied to the detector perpendicular to the 2DEG plane, and the B values in units of Tesla (T) are given on the right-hand side of the figures. The data with the THz irradiation are offset for clarity. (Adapted from Y. Kawano, T. Uchida, and K. Ishibashi, *Appl. Phys. Lett.*, 95, 083123, 2009.)

of the peak shift, which is characteristic of the cyclotron resonance. In addition, from these data, the associated m^* value is derived to be $0.067m_0$, where m_0 is the free electron mass. This value is in good agreement with the cyclotron effective mass for a GaAs-based 2DEG [19]. These facts indicate that the current peak shift of the CNT-SET originates from the THz absorption by the 2DEG.

Let me explain microscopic carrier dynamics associated with the THz detection. It is well known that a 2DEG has a random potential with a typical period of 20–100 nm [20]. It follows that the THz-excited electrons and holes drift in opposite directions through the local electric field gradient due to the random potential [21]. As a result, they are spatially separated from each other. Such separation of electron-hole pairs generates electrical polarization in the 2DEG. This situation is equivalent to application of an effective gate voltage to the CNT-SET, resulting in the current peak shift.

We then measured the temporal trace of the THz signal (the I_{SD} change associated with the current peak shift) as the THz irradiation was cycled on and off. We here used cyclotron radiation [22] from another 2DEG in a different GaAs/AlGaAs chip. The intensity of this THz source can be continuously changed by altering the current

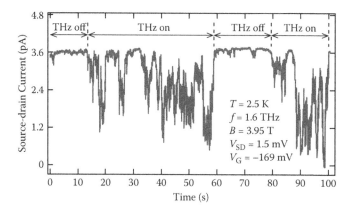

FIGURE 1.8   Time trace of the THz-detected signal (the CNT-SET current I_{SD}) as the THz irradiation is cycled on and off. In this measurement, THz cyclotron radiation emitted from another 2DEG was used with a very low intensity of ~0.1 fW. (Adapted from Y. Kawano, T. Uchida, and K. Ishibashi, *Appl. Phys. Lett.*, 95, 083123, 2009.)

to the 2DEG of the THz source. We reduced its intensity to an extremely low level, on the order of 0.1 fW. We mounted, face to face, the sample and the 2DEG-based THz source in the same superconducting magnet so that the 2DEGs of the sample and the source were both in cyclotron resonance condition. Figure 1.8 displays time response of I_{SD} for an on/off sequence of the THz irradiation at $V_G = -169$ mV, $V_{SD} = 1.5$ mV, $B = 3.95$ T, and $f = 1.6$ THz. This result shows that the CNT-SET current was stable when the THz radiation was off, whereas it repeatedly switched during the THz irradiation. This behavior means that the CNT-SET senses temporal processes of excitation and recombination of THz-excited carriers in the 2DEG.

I note that the CNT-2DEG distance is 120 nm, the relative dielectric constant of GaAs is 13, and the separation length of excited electron-hole pairs is 20–100 nm [20, 21]. Based on these values, the potential change caused by the polarization of a single electron-hole pair is estimated to be 0.4 ~ 10 meV. This value is comparable to the width of the current peak of the CNT-SET (~3 meV). This shows that single-carrier charging via single THz photon absorption in a 2DEG can generate an observable current peak shift and resulting current switching. The data of Figure 1.8 thus demonstrate that the CNT/2DEG device is capable of detecting a few THz photons.

The power of 0.1 fW at 1.6 THz is equivalent to about 10^5 photons per second. This means that the detector does not sense all incident THz photons, which is due to the much smaller detection area than the wavelength of the incident THz wave. I expect that this problem will be resolved by fabricating a device having appropriate antenna-shaped electrodes for the CNT-SET.

The present detector can work at 6 ~ 7 K. This performance eliminates the use of a dilution or ^{3}He refrigerator with low cooling capacity and complex systems. The device can therefore be used in a standard ^{4}He refrigerator or a compact cryo-free mechanical refrigerator with much higher cooling capacity and much greater ease of use. This advantageous capability makes it possible to extend the usable range of ultra-sensitive THz measurements.

1.3   THz IMAGING

Enhancing spatial resolution of THz imaging is one of the central issues in THz technology. Nevertheless, this is a formidable task, because basic components necessary for obtaining high resolution, such as high-transmission wave line and a highly sensitive detector, have not been well established compared to other frequency regions.

There are generally two methods for realizing high resolution in optical imaging: solid immersion lens and near-field imaging technique. I explain the applications of the techniques in the THz imaging below.

1.3.1   Solid Immersion Lens

In a standard optical imaging system like the use of a lens, spatial resolution of an optical image is proportional to a wavelength of the light. The solid immersion lens utilizes the fact that a wavelength in a material with high dielectric constant is reduced compared to that in vacuum. Therefore, a lens with a large refractive index leads to high resolution.

Figure 1.9 depicts an example of THz imaging based on the solid immersion lens [23]. A hyperhemispherical lens made of Si is in contact with the back surface of a sample. The sample is a 2DEG in a GaAs/AlGaAs heterostructure. A THz emission from the 2DEG is radiated as a consequence of inter-Landau-level transitions of electrons under magnetic field, i.e., cyclotron emission (CE) [22]. The focal point of the lens is designed to be on the 2DEG layer of the sample. The CE from the focal point is collimated via the lens and is guided through a metallic light pipe to a THz detector. Off-axis light is filtered by a black polyethylene pipe. The sample is scanned relative to the lens by use of an *X-Y* translation stage, while the optical system including the detector is stationary. The relatively large refractive index of GaAs and Si ($n \sim 3.4$) allows a resolution much higher than that obtained with a remote lens system.

Kawano and Komiyama [23] reported that for the solid immersion lens system, the resolution is about 50 μm at 130 μm, the wavelength of CE in vacuum. This value is improved by a factor of 6, compared to the resolution (300 μm) for the remote Si lens.

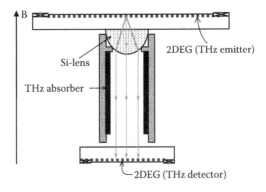

FIGURE 1.9   THz microscope based on a solid immersion lens.

1.3.2 NEAR-FIELD IMAGING

Though the solid immersion lens system enables relatively high resolution, the resolution is still determined by the diffraction limit. A powerful technique for overcoming the diffraction limit is near-field imaging [24]. When an electromagnetic wave is illuminated onto an aperture smaller than a wavelength, a localized evanescent field (near field) is generated just behind the aperture. The size of the evanescent field is determined by the aperture size and not the wavelength. By illuminating or detecting the evanescent field, it is possible to get a resolution beyond the diffraction limit.

Conventional systems for near-field imaging have an aperture type and an apertureless (scattering) type. In the visible and near-infrared regions, either a tapered, metal-coated optical fiber (aperture type) [25] or a metal tip (aperture-less type) [26] is used. In the microwave region, either a sharpened waveguide (aperture type) [27] or a coaxial cable (aperture-less type) [28] is used. Since the intensity of the evanescent field is very weak, the realization of near-field imaging needs a highly sensitive detection scheme, such as a high-transmission wave line and highly sensitive detector.

1.3.2.1 Near-Field Imaging in the THz Region

Several methods for near-field imaging in the THz region have been presented. In the aperture type, a metal hole was used to obtain a resolution better than $\lambda/4$ [29]. This method, however, has the drawback of low wave transmission through the small aperture, which requires detecting very weak waves. Moreover, the spatial resolution is not so high. As the aperture-less type, a sharpened antenna [30], a metal tip [31], and a cantilever [32] were reported. Although the aperture-less technique allows high spatial resolution, it has the problem of separation from a far-field component of an incident wave, which generates a large background signal. For this reason, in most instances the probe position is modulated and synchronously detected signal is measured. However, this scheme makes the whole system and its operation complicated.

At present stage, near-field THz imaging has not been fully established, and several other techniques are proposed and attempted. This is still an ongoing issue.

1.3.2.2 Near-Field THz Imaging on One Chip

I here explain unique near-field THz imaging—on-chip THz imaging with an integrated semiconductor device [33, 34]. The problems of other techniques were low efficiency of near-field detection and complicated systems. The new integrated device enables near-field THz imaging in a much simpler manner. As shown in Figure 1.10, an aperture with an 8 μm diameter and a planar probe were deposited on a surface of a GaAs/AlGaAs heterostructure chip. The aperture and the probe are insulated by a 50 nm thick SiO_2 layer. The GaAs chip has an electron mobility of 18 m^2/Vs and a sheet electron density of 4.4×10 m^{-2} at 77 K. The two ohmic contacts were extended to the side surfaces of the chip, to each of which an electrical wire was attached. A 2DEG, located 60 nm below the chip surface, works as a THz detector. In this device structure, all components—an aperture, a probe, and a detector—are integrated on one GaAs/AlGaAs chip. This scheme eliminates any optical and mechanical alignments between each component, leading to an easy-to-use and robust system.

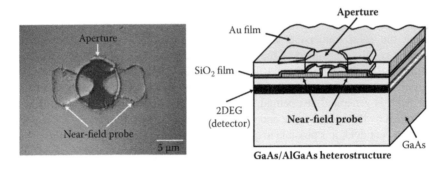

FIGURE 1.10   Photograph of the near-field THz imaging device and its schematic view.

An advantage of this device structure is that the presence of the planar probe changes the distribution of the evanescent field, enhancing the coupling of the evanescent field to the 2DEG detector. This is expected to lead to an increase in the detection sensitivity. Moreover, the 2DEG detector is not affected by the far-field wave owing to the close vicinity to the aperture and the probe, allowing the detection of the evanescent field alone.

Figure 1.11 shows calculations of THz electric field distributions near the aperture using a finite element method. Compared are the device of the aperture alone and that of the aperture plus the probe, where the aperture diameters are the same. For the case of the aperture alone, the electric field is localized close to the aperture. This explains the reason why the conventional aperture-based technique suffers from low transmission through the aperture. On the other hand, when the probe is present just behind the aperture, the electric field extends into the interior region of the GaAs substrate. This result indicates that the presence of the probe changes the distribution profile of the evanescent field, enhancing the coupling between the evanescent field and the 2DEG detector.

In order to confirm experimentally the above calculations, we measured the THz transmission distribution by scanning the device across a sample. In the measurements, the THz gas laser pumped by the CO_2 gas laser was used as a THz source. The THz radiation was chopped at 16 Hz and the amplitude modulation of the detected signal (the voltage change of the 2DEG) was measured with a lock-in amplifier. The THz near-field device relative to the sample was scanned with a translation stage based on a piezoelectric stick-slip motion. The sample is made up of a THz transparent substrate, the surface of which is covered at regular intervals by THz opaque Au films. The widths of THz opaque and transparent regions across the scan direction are 80 and 50 µm, respectively. Two types of near-field THz devices were used: the aperture plus the probe and the aperture alone. As displayed in Figure 1.12(a), in the former case, a clear signal is visible, whereas in the latter case, no signal is observed. This feature does not depend on the wavelength of the incident THz wave. The enhancement ratio of the signal amplitude regarding comparison between the two devices was found to be 41 for $\lambda = 118.8$ µm and 67 for $\lambda = 214.6$ µm. These results clearly demonstrate that the distribution of the THz evanescent field is largely enhanced due to the presence of the probe, leading to improvement in the detection sensitivity.

Aperture Alone

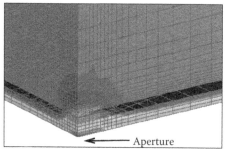

Aperture + Probe

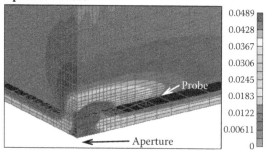

FIGURE 1.11 Calculations of THz electric field distributions in the vicinity of the aperture for the device with the aperture and probe (upper panel) and for the device with the aperture alone (lower panel).

The decay curves of Figure 1.12(b) show that the near-field THz device has a spatial resolution of 9 μm. The important points of the data are that the resolution does not depend on the wavelength of the THz wave, and that it is far beyond the diffraction limit λ/2 (107.3 μm) for the wavelength of λ = 214.6 μm. These features are characteristic of the near-field effect. The present device hence properly functions as a near-field THz imaging detector. Using this device, a high-resolution image of THz CE distribution in another 2DEG has been obtained [35].

In general, when one improves spatial resolution, one encounters the problem that it is necessary to largely enhance detection sensitivity. This is because the intensity of the wave to be detected is strongly reduced as the sensing area becomes small. This situation compels one to produce a scheme of highly sensitive detection in a tiny region. I expect that one promising approach for tackling this problem in the THz region will be to use the CNT/2DEG THz detector [13], described in the previous section. Compared to the 2DEG detector, the CNT/2DEG detector exhibits much higher sensitivity in a much smaller sensing area of submicrometer. When this detector is integrated with an aperture and a probe as shown in Figure 1.10, the device will enable THz imaging with ultra-high sensitivity and a sub-μm resolution.

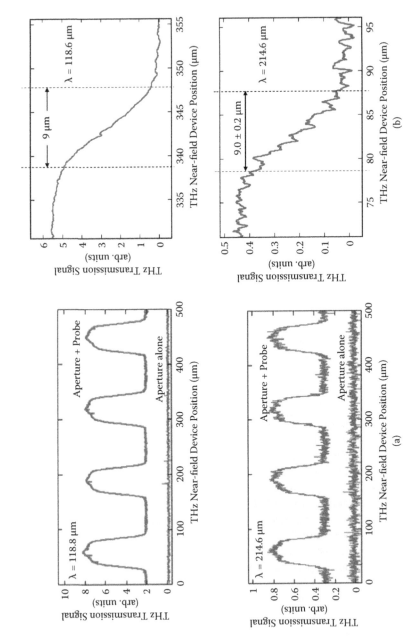

FIGURE 1.12 (a) THz transmission signal versus the position of the near-field THz detector. (b) Decay curves of the signals in (a).

1.4 THz APPLICATION

1.4.1 APPLICATIONS TO MATERIALS RESEARCHES

The photon energy and the period of the THz wave with 1 THz are about 4 meV and 1 ps, respectively. In these energy and time regions, there are many important materials and their physical properties, such as energy gap of superconductor, impurity level of semiconductor, phonon energy, and Landau level separation. THz spectroscopy and imaging can thus be exploited as a powerful tool for investigating and characterizing various materials.

There are a lot of THz applications to materials researches. Shikii et al. [36] reported mapping of supercurrent distributions in a high-temperature superconductor by means of the THz imaging technique. THz-TDS is a very powerful tool for investigating electron dynamics in materials. This technique provides two kinds of information: real-time dynamics and frequency spectra. With this method, a lot of important information was successfully derived and the related electronic properties were clarified. For example, reported were THz conductivity in an MgB_2 superconductor [37], Bloch oscillation in a semiconductor superlattice [38], ballistic electron motion in a CNT [39], and dynamics of a spin-density-wave gap in a quasi-1D organic conductor [40].

1.4.2 APPLICATIONS TO SEMICONDUCTOR

1.4.2.1 Imaging Energy Dissipation in 2DEG

When the 2DEG is subject to a perpendicular magnetic field, it exhibits dramatic properties known as quantum Hall effect (QHE) [41, 42]. In the QHE regime, the longitudinal resistance vanishes and the Hall resistance is precisely quantized. The basis of the QHE is the formation of the Landau level, which originates from the quantization of electron cyclotron motion. Studying electron distribution of an excited Landau level (related energy-dissipation distribution) is one of the critical issues regarding the QHE. The reason is as follows: Earlier works reported that the electron relaxation from excited Landau level involves a slow physical process, the typical timescale being as long as 10–100 ns [43]. Moreover, the equilibrium length of the excited electrons L_E was shown to reach a macroscopic value of ~0.1 mm [44, 45]. This leads us to expect that spatial distribution of the excited electrons will be strikingly different from that of the ground-state electrons and will drastically depend on sample dimensions, in relation with L_E. However, this issue has not been satisfactorily clarified.

Several imaging techniques have been applied to obtain information on dissipation distributions in the QHE systems. By using the fountain pressure effect of superfluid liquid helium, Klaß et al. demonstrated that in the QHE state, dissipation takes place almost totally at the diagonally opposite electron entry and exit corners of the sample (hot spots) [46]. A similar conclusion has been derived from the experiments of Russell et al., which applied a local bolometry technique [47]. Unfortunately, since these earlier experiments probe the lattice temperature related to Joule heating, they do not exclusively observe the distribution within the 2DEG, i.e., the effective temperature of the excited electrons. Thus, until now no experimental techniques were available for imaging only the dissipation distribution in the 2DEG.

In contrast, since CE is radiated with the transition of electrons from higher Landau levels to lower Landau levels, mapping the CE intensity with the THz microscope provides a direct probe of the excited electrons. In the following, I introduce investigations on spatial properties of the 2DEGs by use of THz-CE imaging.

1.4.2.2   Separate Imaging of Ground and Excited Levels

In order to enable separate imaging of the ground-state electrons and the excited-state electrons, we have developed a combined system of an electrometer with a THz microscope [48]. In the electrometer technique, we are able to map voltage distributions [49, 50]. Hence, comparing the two kinds of the mapped data (voltage and CE) allows us to distinguish the contributions of the ground-state and excited-state electrons to the generation of the voltage.

Figure 1.13 depicts the setup of the combined imaging system. An electrometer (small Hall bar) and a hyperhemisphere Si lens are in contact with the front and back surfaces of the sample, respectively. A THz detector is mounted at the bottom of the system. The electrometer, the sample, and the THz detector were fabricated on GaAs/AlGaAs heterostructure wafers having the 2DEGs 0.1 μm beneath the crystal surfaces. In imaging experiments, the sample alone was moved with an X-Y stage, while the electrometer and the optical system were spatially fixed. The whole system was directly immersed into liquid ^{4}He and was subject to a perpendicular magnetic field B.

The detection mechanism of the electrometer is as follows: As shown in Figure 1.14, the distance between the two 2DEG layers is very short, and consequently they are capacitively coupled. It follows that when a current I_s is passed through the sample, excess charges ΔQ are induced into the 2DEGs, according to $CV(x,y) = \Delta Q$. Here, $V(x,y)$ is a local voltage in the sample right below the electrometer, and

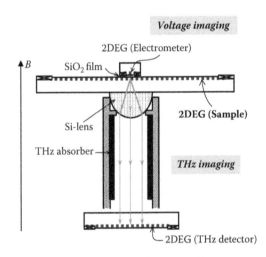

FIGURE 1.13   Combined imaging system of the electrometer and the THz microscope.

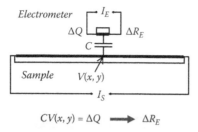

$$CV(x, y) = \Delta Q \longrightarrow \Delta R_E$$

FIGURE 1.14 Equivalent circuit of the scanning electrometer system and the detection mechanism. (Adapted from Y. Kawano, and T. Okamoto, *Phys. Rev.*, B 70, 081308(R), 2004.)

C is the capacitance between the two 2DEG layers. The generation of ΔQ leads to the change, ΔR_E, in the longitudinal resistance of the electrometer. By translating the electrometer over the sample surface and measuring $\Delta R_E \times I_E$, we are able to image 2D distributions of $V(x,y)$ in the sample (I_E is the bias current for the electrometer). The electrometer was processed into a Hall bar geometry from a GaAs/AlGaAs heterostructure with an electron mobility $\mu_H = 18$ m^2/Vs and a sheet electron density $n_s = 2.2 \times 10$ m^{15} m^{-2} at 4.2 K. The device has a 2DEG channel with a length $L = 200$ μm and a width $W = 1.5$ μm, and the two voltage probes with an interval of 2.3 μm. The spatial resolution of this system is about 2 μm, which is determined by the sensing 2DEG area of the electrometer.

The sample studied has a Hall bar geometry with $L = 2.8$ mm and $W = 1$ mm, which is fabricated from a GaAs/AlGaAs heterostructure with $\mu_H = 53$ m^2/Vs and $n_s = 2.4 \times 10^{15}$ m^{-2} at 4.2 K. In imaging measurements, the voltage $V(x,y)$ and the CE intensity $V_{CE}(x,y)$ were simultaneously obtained with two lock-in amplifiers, where a 30 Hz rectangular-wave current alternating between zero and a given finite value I_s was passed through the sample. All measurements were performed at 4.2 K.

Figure 1.15 displays images of the longitudinal voltage V_{xx} and V_{CE} at three current levels, $I = 20$, 70, and 140 μA, and at $B = 5.75$ T (Landau level filling factor, 1.74, of the sample). We obtained the $V_{xx}(x,y)$ data by differating $V(x,y)$ over a distance of 50 μm in the direction of the length of the sample: $V_{xx}(x,y) = V(x + \Delta x,y) - V(x,y)$ with $\Delta x = 50$ μm. The 20 μA data show that V_{xx} is seen over the whole 2DEG area, whereas CE occurs only in the two diagonally opposite corners. This directly indicates that the observed V_{xx} (except at the two spots) arises from scattering of the ground-state electrons. It is known that the CE at the two spots takes place via tunneling injection of nonequilibrium electrons from the source contact and inter-Landau-level tunneling of electrons near the drain contact [22]. The V_{xx} data at $I = 70$ and 140 μA show that an additional V_{xx} is generated around the source contact, and with increasing the current, the area covered by the additional V_{xx} expands toward the interior 2DEG region. Comparing with the V_{CE} data reveals that a strong CE is radiated in the similar region, indicating that V_{xx} observed on the source contact side mostly originates from scattering of the excited electrons. V_{xx} is also visible over the whole 2DEG region, whereas CE remains absent (except in the two diagonally opposite corners). This situation is similar to that observed at $I = 20$ μA, showing again that V_{xx} associated with the ground-state electrons is generated over

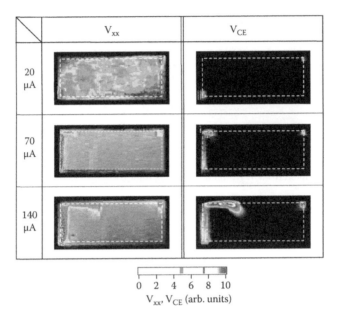

FIGURE 1.15 Distribution images of the longitudinal voltage V_{xx} (left) and the CE intensity V_{CE} (right) at three current levels, $I = 20$, 70, and 140 μA. The white broken lines denote the 2DEG boundaries of the sample. In the V_{xx} (V_{CE}), for clarity, the signals at $I = 20$ and 70 μA are magnified, respectively, by factors of 15 (18) and 9 (3) compared to that at $I = 140$ μA.

the entire 2DEG area. We confirmed that the pattern of these features systematically changes upon reversing B and I, and that they are also observed for other Landau level filling factors and for other samples fabricated from different wafers. This indicates that they originate from an intrinsic nature of the 2DEG and are not due to inhomogeneity of the electron density.

Let me explain the origin of the difference in the distributions between the ground-state electrons and the excited electrons. The summary of the experimental findings is that as schematically shown in Figure 1.16, the scattering distribution of the ground-state electrons spreads over the entire 2DEG region, whereas that of the excited electrons is localized in the two corners and around the source contact. The result indicates a large difference in the scattering processes and their characteristic lengths. The equilibrium length of excited electrons was shown to be a very long scale of 10–300 μm [44, 45]. On the other hand, although there is no clear experimental information about the ground-state electrons, one can think that the scattering rate will be much higher because the scattering does not require much energy. The main scattering mechanism of the ground-state electrons is possibly impurity scattering originating from ionized donors in the AlGaAs layer. The characteristic scattering length of the ground-state electrons can be regarded as a period of the ionized impurities, 0.05–0.3 μm, which is much shorter than L_E. This explains the contrasting situation regarding spatial distributions of ground- and excited-state electrons. I will discuss the electron transport process in more detail below.

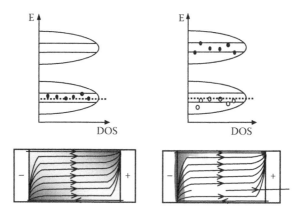

FIGURE 1.16 Sketch of scattering distributions of the ground-state electrons (left) and of the excited-state electrons (right).

1.4.2.3 Macroscopic Channel Size Effect of THz Image

The experimental data in the above showed that the macroscopically long value of L_E results in the asymmetric distribution of the excited electrons. From this fact, the following interesting phenomenon is expected: spatial distribution of the excited electrons will drastically depend on sample dimensions, in relation with L_E. We have addressed this issue by studying the channel size dependence of the CE distribution by use of THz imaging [51].

The experimental setup of the THz microscope is similar to that shown in Figure 1.9. Three samples with different widths of $W = 1.2, 0.3$, and 0.02 mm (length $L = 4$ mm) were investigated, all of which were fabricated from the same GaAs/AlGaAs heterostructure crystal with $\mu_H = 62$ m^2/Vs and $n_s = 2.4 \times 10^{15}$ m^{-2} at 4.2 K. In CE measurements, the detected signal was recorded with a lock-in amplifier, where 20 Hz rectangular-wave currents alternating between zero and a given finite value I were transmitted through the sample. All the measurements were performed at 4.2 K.

Figure 1.17 shows comparison of the CE images for the three samples. In the widest sample ($W = 1.2$ mm), CE takes place mostly in the two diagonally opposite corners and in a localized region near the source contact. In the data for $W = 0.3$ mm, CE expands into the more interior 2DEG region. Furthermore, when $W = 0.02$ mm, the CE takes place over the entire region of the interior 2DEG channel. We confirmed that the pattern of these features systematically changed upon reversing B and I, and that the similar CE images were also observed for other samples fabricated from different GaAs/AlGaAs wafers.

The significant change with W observed shows the validity of the above expectation: spatial distribution of the excited electrons is considerably influenced by the very long length scale of L_E. This means that the electrons travel the macroscopic distance L_E to lead to an appreciable generation of CE. Based on this, I explain the transport process of the excited electrons below.

In the QHE devices, it is known that high electric field is concentrated in the two diagonally opposite corners, which are also electron entry and exit corners [52].

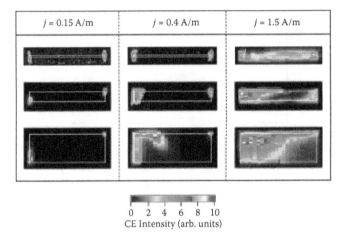

$j = 0.15$ A/m	$j = 0.4$ A/m	$j = 1.5$ A/m

0   2   4   6   8 10
CE Intensity (arb. units)

FIGURE 1.17 CE images at three current densities, $j = 0.15$, 0.4, and 1.5 A/m, and for the three samples with different channel widths, $W = 20$, 300, and 1200 μm. The white solid lines denote the 2DEG boundaries of the samples. For clarity, the signals are magnified by factors of 1300 (20, 0.15), 360 (20, 0.4), 23 (20, 1.5), 70 (300, 0.15), 18 (300, 0.4), 3 (300, 1.5), 14 (1200, 0.15), and 5 (1200, 0.4), respectively, compared to that for (1200, 1.5), where (A, B) represents (W in μm, j in A/m). (Adapted from Y. Kawano, and T. Okamoto, *Phys. Rev. Lett.*, 95, 166801, 2005.)

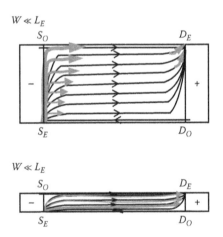

FIGURE 1.18 Schematic view of trajectories (orange lines) of excited electrons in the QHE device for a wide Hall bar with $W \gg L_E$ (upper panel) and a narrow Hall bar with $W \ll L_E$ (lower panel).

It follows that as the current increases, electron excitation into a higher Landau level starts in the vicinity of the two current contacts. On the source contact side, CE begins to take place in the electron entry corner S_E (Figure 1.18). In $W \gg L_E$, though the electric field decreases away from S_E, the excitation process can continue toward the opposite corner (S_O) over the distance L_E. As the applied electric field becomes higher, like 0.4 and 1.5 A/m in the sample with $W = 1.2$ mm, the area where such excitation occurs further expands through S_O toward the interior region of the 2DEG channel. Similar excitation process takes place along the boundary of the drain contact. In this region, on the other hand, electrons travel from the opposite corner (D_O) to the electron exit corner (D_E), which is opposite to that on the side of the source contact. As a result, though electron excitation starts around D_O, CE does not appreciably occur there and mostly occurs at D_E.

In contrast, when W is smaller than L_E, CE is not generated in the vicinity of the source contact. This is because in such narrow devices, the electron excitation can take place only as the electrons traverse the Hall bar along the length direction of the 2DEG channel. The CE thus occurs over the whole region of the interior 2DEG channel, which is in strong contrast to the CE distribution for the devices with $W \gg L_E$.

1.5 CONCLUSION AND OUTLOOK

In this chapter, I have presented novel THz sensing and imaging devices based on nano-electronic materials and devices: 2D semiconductor and CNT. These detectors enable THz wave sensing with high sensitivity and high spatial resolution. I also introduce applications of THz measurements to materials researches. The experimental data have successfully revealed spatial properties of electrons in the 2DEG, which have not been revealed by conventional transport measurements.

The next key devices for future THz technology will be THz cameras, THz video, and THz communication. When these devices are realized in portable size, THz technology will widely spread into our life. For this purpose, achieving higher detection sensitivity and higher power generation will increasingly become important tasks. The development should require further breakthroughs in materials, device structure, operation principles, etc. I anticipate that nanotechnology will still be one of the keys for enhancing performance of THz technology.

Regarding our next target, I plan to create a THz photon counting video, which is a strongly desired device in many fields. When many CNT-based THz photon detectors [13] are integrated in a two-dimensional configuration [53, 54], the resulting device will work as a THz video for real-time, high-resolution THz imaging.

REFERENCES

1. B. Ferguson and X.-C. Zhang, Materials for terahertz science and technology, *Nat. Mater.* 1, 26 (2002).
2. M. Tonouchi, Cutting-edge terahertz technology, *Nat. Photonics* 1, 97 (2007).
3. J. Wei, D. Olaya, B. S. Karasik, S. V. Pereverzev, A. V. Sergeev, and M. E. Gershenson, Ultrasensitive hot-electron nanobolometers for terahertz astrophysics, *Nat. Nanotechnol.* 3, 496 (2008).
4. S. Ariyoshi, C. Otani, A. Dobroiu, H. Matsuo, H. Sato, T. Taino, K. Kawase, and H. Shimizu, Terahertz imaging with a direct detector based on superconducting tunnel junctions, *Appl. Phys. Lett.* 88, 203503 (2006).
5. S. Komiyama, O. Astafiev, V. Antonov, H. Hirai, and T. Kutsuwa, A single-photon detector in the far-infrared range, *Nature* 403, 405 (2000).

6. P. Kleinschmidt, S. Giblin, A. Tzalenchuk, H. Hashiba, V. Antonov, and S. Komiyama, Sensitive detector for a passive terahertz imager, *J. Appl. Phys.* 99, 114504 (2006).

7. R. Saito, G. Dresselhaus, and M. S. Dresselhaus, *Physical properties of carbon nanotubes*, Imperial College Press, London, 1998.

8. P. W. Barone, S. Baik, D. A. Heller, and M. S. Strano, Near-infrared optical sensors based on single-walled carbon nanotubes, *Nat. Mater.* 4, 86 (2005).

9. M. E. Itkis, F. Borondics, A. Yu, and R. C. Haddon, Bolometric infrared photoresponse of suspended single-walled carbon nanotube films, *Science* 312, 413 (2006).

10. T. Fuse, Y. Kawano, M. Suzuki, Y. Aoyagi, and K. Ishibashi, Coulomb peak shifts under terahertz-wave irradiation in carbon nanotube single-electron transistors, *Appl. Phys. Lett.* 90, 013119 (2007).

11. K. Fu, R. Zannoni, C. Chan, S. H. Adams, J. Nicholson, E. Polizzi, and K. S. Yngvesson, Terahertz detection in single wall carbon nanotubes, *Appl. Phys. Lett.* 92, 033105 (2008).

12. Y. Kawano, T. Fuse, S. Toyokawa, T. Uchida, and K. Ishibashi, Terahertz photon-assisted tunneling in carbon nanotube quantum dots, *J. Appl. Phys.* 103, 034307 (2008).

13. Y. Kawano, T. Uchida, and K. Ishibashi, Terahertz sensing with a carbon nanotube/two-dimensional electron gas hybrid transistor, *Appl. Phys. Lett.* 95, 083123 (2009).

14. P. K. Tien and J. P. Gordon, Multiphoton process observed in the interaction of microwave fields with the tunneling between superconductor films, *Phys. Rev.* 129, 647 (1963).

15. J. M. Hergenrother, M. T. Tuominena, J. G. Lua, D. C. Ralpha, and M. Tinkham, Charge transport and photon-assisted tunneling in the NSN single-electron transistor, *Physica B* 203, (1994) 327.

16. B. Leone, J. R. Gao, T. M. Klapwijk, B. D. Jackson, W. M. Laauwen, and G. de Lange, Electron heating by photon-assisted tunneling in niobium terahertz mixers with integrated niobium titanium nitride striplines, *Appl. Phys. Lett.* 78, 1616 (2001).

17. T. H. Oosterkamp, L. P. Kouwenhoven, A. E. A. Koolen, N. C. van der Vaart, and C. J. P. M. Harmans, Photon sidebands of the ground state and first excited state of a quantum dot, *Phys. Rev. Lett.* 78, 1536 (1997).

18. T. H. Oosterkamp, T. Fujisawa, W. G. van der Wiel, K. Ishibashi, R. V. Hijman, S. Tarucha, and L. P. Kouwenhoven, Microwave spectroscopy of a quantum-dot molecule, *Nature* 395, 873 (1998).

19. F. Thiele, U. Merkt, J. P. Kotthaus, G. Lommer, F. Malcher, U. Rossler, and G. Weimann, Cyclotron masses in n-GaAs/Ga1-$_x$Al$_x$As heterojunctions, *Solid State Commun.* 62, 841 (1987).

20. S. H. Tessmer, P. I. Glicofridis, R. C. Ashoori, L. S. Levitov, and M. R. Melloch, Subsurface charge accumulation imaging of a quantum Hall liquid, *Nature* 392, 51 (1998).

21. Y. Kawano, Y. Hisanaga, H. Takenouchi, and S. Komiyama, Highly sensitive and tunable detection of far-infrared radiation by quantum Hall devices, *J. Appl. Phys.* 89, (2001) 4037.

22. Y. Kawano, Y. Hisanaga, and S. Komiyama, Cyclotron emission from quantized Hall devices: Injection of nonequilibrium electrons from contacts, *Phys. Rev. B* 59, 12537 (1999).

23. Y. Kawano and S. Komiyama, Spatial distribution of non-equilibrium electrons in quantum Hall devices: Imaging via cyclotron emission, *Phys. Rev. B* 68, 085328 (2003).

24. M. Ohtsu, ed., *Near-field nano/atom optics and technology*, Springer-Verlag, Berlin, 1998.

25. T. Saiki, S. Mononobe, M. Ohtsu, N. Saito, and J. Kusano, Tailoring a high-transmission fiber probe for photon scanning tunneling microscope, *Appl. Phys. Lett.* 68, 2612 (1996).

26. F. Zenhausern, Y. Martin, and H. K. Wickramasinghe, Scanning interferometric apertureless microscopy: Optical imaging at 10 angstrom resolution, *Science* 269, 1083 (1995).

27. W. C. Symons III, K. W. Whites, and R. A. Lodder, Theoretical and experimental characterization of a near-field scanning microwave microscope (NSMM), *IEEE Trans. Microwave Theory Tech.* 51, 91 (2003).

28. M. Tabib-Azar and Y. Wang, Design and fabrication of scanning near-field microwave probes compatible with atomic force microscopy to image embedded nanostructures, *IEEE Trans. Microwave Theory Tech.* 52, 971 (2004).

29. S. Hunsche, M. Koch, I. Brener, and M. C. Nuss, THz near-field imaging, *Opt. Commun.* 150, 22 (1998).

30. N. C. J. van der Valk and P. C. M. Planken, Electro-optic detection of subwavelength terahertz spot sizes in the near field of a metal tip, *Appl. Phys. Lett.* 81, 1558 (2002).

31. H. T. Chen, R. Kersting, and G. C. Cho, Terahertz imaging with nanometer resolution, *Appl. Phys. Lett.* 83, 3009 (2003).

32. A. J. Huber, F. Keilmann, J. Wittborn, J. Aizpurua, and R. Hillenbrand, Terahertz near-field nanoscopy of mobile carriers in single semiconductor nanodevices, *Nano Lett.* 8, 3766 (2008).

33. Y. Kawano and K. Ishibashi, An on-chip near-field terahertz probe and detector, *Nat. Photonics* 2, 618 (2008).

34. Y. Kawano, Terahertz detectors: Quantum dots enable integrated terahertz imager, *Laser Focus World* 45(7), 45–47, 50 (2009).

35. Y. Kawano and K. Ishibashi, On-chip near-field terahertz detection based on a two-dimensional electron gas, *Physica E*, 42, 1188 (2010).

36. S. Shikii, T. Kondo, M. Yamashita, M. Tonouchi, M. Hangyo, M. Tani, and K. Sakai, Observation of supercurrent distribution in $YBa_2Cu_3C>_{7-\delta}$ thin films using THz radiation excited with femtosecond laser pulses, *Appl. Phys. Lett.* 74 (1999) 1317.

37. R. A. Kaindl, M. A. Carnahan, J. Orenstein, D. S. Chemla, H. M. Christen, H.-Y. Zhai, M. Paranthaman, and D. H. Lowndes, Far-infrared optical conductivity gap in superconducting MgB2 films, *Phys. Rev. Lett.* 88 (2001) 027003.

38. N. Sekine and K. Hirakawa, Dispersive terahertz gain of non-classical oscillator: Bloch oscillation in semiconductor superlattices, *Phys. Rev. Lett.* 94, 057408 (2005).

39. Z. Zhong, N. M. Gabor, J. E. Sharping, A. L. Gaeta, and P. L. McEuen, Terahertz time-domain measurement of ballistic electron resonance in a single-walled carbon nanotube, *Nat. Nanotechnol.* 3 (2008) 201.

40. S. Watanabe, R. Kondo, S. Kagoshima, and R. Shimano, Observation of ultrafast photoinduced closing and recovery of the spin-density-wave gap in $(TMTSF)_2PF_6$, *Phys. Rev. B* 80, 220408(R) (2009).

41. K. von Klitzing, G. Dorda, and M. Pepper, New method for high-accuracy determination of the fine-structure constant based on quantized Hall resistance, *Phys. Rev. Lett.* 45, 494 (1980).

42. D. C. Tsui, H. L. Stormer, and A. C. Gossard, Two-dimensional magnetotransport in the extreme quantum limit, *Phys. Rev. Lett.* 48, 1559 (1982).

43. B. R. A. Neves, N. Mori, P. H. Beton, L. Eaves, J. Wang, and M. Henini, Landau-level populations and slow energy relaxation of a two-dimensional electron gas probed by tunneling spectroscopy, *Phys. Rev. B* 52, 4666 (1995).

44. S. Komiyama, Y. Kawaguchi, T. Osada, and Y. Shiraki, Evidence of nonlocal breakdown of the integer quantum Hall effect, *Phys. Rev. Lett.* 77, 558 (1996).

45. I. I. Kaya, G. Nachtwei, K. von Klitzing, and K. Eberl, Spatially resolved monitoring of the evolution of the breakdown of the quantum Hall effect: Direct observation of inter-Landau-level tunneling, *Europhys. Lett.* 46, 62 (1999).

46. U. Klaß, W. Dietsche, K. von Klitzing, and K. Ploog, Imaging of the dissipation in quantum-Hall-effect experiments, *Z. Phys. B* 82, 351 (1991).

47. P. A. Russell, F. F. Ouali, N. P. Hewett, and L. J. Challis, Power dissipation in the quantum Hall regime, *Surface Sci.* 229, 54 (1990).

48. Y. Kawano and T. Okamoto, Imaging of intra- and inter-Landau-level scattering in quantum Hall systems, *Phys. Rev. B* 70, 081308(R) (2004).

49. Y. Kawano and T. Okamoto, Scanning electrometer using the capacitive coupling in quantum Hall effect devices, *Appl. Phys. Lett.* 84, 1111 (2004).

50. Y. Kawano and T. Okamoto, Noise-voltage mapping by a quantum-Hall electrometer, *Appl. Phys. Lett.* 87, 252108 (2005).

51. Y. Kawano and T. Okamoto, Macroscopic channel-size effect of nonequilibrium electron distributions in quantum Hall conductors, *Phys. Rev. Lett.* 95, 166801 (2005).

52. J. Wakabayashi and S. Kawaji, Hall effect in silicon MOS inversion layers under strong magnetic fields, *J. Phys. Soc. Jpn.* 44, 1839 (1978).

53. N. R. Franklin, Q. Wang, T. W. Tombler, A. Javey, M. Shim, and H. Dai, Integration of suspended carbon nanotube arrays into electronic devices and electromechanical systems, *Appl. Phys. Lett.* 81, 913 (2002).

54. H. Tabata, M. Shimizu, and K. Ishibashi, Fabrication of single electron transistors using transfer-printed aligned single walled carbon nanotubes array, *Appl. Phys. Lett.* 95, 113107 (2009).

2 Ultimate Fully Depleted (FD) SOI Multigate MOSFETs and Multibarrier Boosted Gate Resonant Tunneling FETs for a New High-Performance Low-Power Paradigm

Aryan Afzalian

CONTENTS

As transistors are scaled down in the nanoscale regime, quantum effects are playing a crucial role in device performance and parameters. Also, scaling alone is not sufficient to achieve performance improvement, and new boosters and device concepts are needed. For instance, the trade-off between power and performance in electronics is one of the most limiting factors to push further technology scaling and development. With scaling, the reduction of supply voltage to keep power density under control [1–3], the rise of source and drain resistance due to film thickness reduction in order to keep good electrostatic control [3], and finally, source-drain (SD) tunneling that degrades subthreshold slope and increases leakage of transistors below 10 nm [3, 4] are major roadblocks that degrade on- and off-current, and I_{ON}/I_{OFF} ratios, and therefore the power-delay trade-off of transistors. I_{ON}/I_{OFF} ratios and slope characteristics of transistors depend on the gate-to-channel coupling and carrier statistics that dictate the way carriers are made available to drive a current when increasing V_G. By reducing film thickness and increasing the number of gates, ultra-thin film multigate silicon-on-insulator (SOI) architectures have better electrostatic control and can achieve near-ideal subthreshold slope and improved I_{ON}/I_{OFF} ratio over more conventional bulk Si single-gate architectures. Supposing ideal gate coupling, however, when varying the gate voltage, the current varies at a rate dictated by Fermi-Dirac statistics only. This is governed by the gate-controlled single-barrier paradigm on which present field-effect transistors (FETs) are based. In a standard transistor, there is only one barrier from channel to source, and the density of state close to the top of the channel barrier is about constant with V_G [5]. As a result, the current increase is exponential below threshold with an optimal minimal inverse subthreshold slope (SS) of kT/q.log 10, i.e., about 60 mV/decade at $T = 300$ K. Above threshold, when the channel barrier passes below the source Fermi level, E_{FS}, enabling the source highly occupied states to drive a significant current density, and thus good delay performance, the current increase is much slower and the inverse slope reaches much higher values.

We have recently shown the possibility of achieving better slopes than those dictated by Fermi-Dirac, both in subthreshold and above threshold, together with high on-current, by using a Si multibarrier boosted complementary metal-oxide-semiconductor (CMOS) transistor, the gate modulated resonant tunneling (RT) FET [5, 6]. It is a metal-oxide-semiconductor field-effect transistor (MOSFET) boosted with additional tunnel barriers (TBs) (i.e., barriers of a few nanometers width and less than 10 nm) near the gate edge(s) and under electrostatic control of the gate that creates additional longitudinal confinement in the device. Such TBs can be created, for instance, in a planar technology from a local reduction, or constriction, of the device cross section, resulting from a local oxidation that can be well controlled [7], or from Schottky barriers and dopant segregation techniques [6, 8] in this case allowing for steep slope and low source and drain resistance. RT-FETs have also been shown to be immune to the source-drain tunneling problem that further degrades I_{ON}/I_{OFF} ratios in standard devices for channel lengths below 10 nm.

In this chapter, we investigate and compare the performances of ultimate SOI multigate nanowire FETs with channel lengths of about 10 nm and below through nonequilibrium Green's function (NEGF) quantum simulations, within both ballistic and scattering self-consistent Born approximations. Both standard single-barrier devices and the new multibarrier boosted architecture are compared. In Section 2.1, the simulation algorithm is reviewed. In Section 2.2, quantum effects and their impact on the gate coupling optimization of ultra-scaled nanowires are described by optimizing I_{ON}/I_{OFF} ratios vs. the cross section size in a 10 nm gate-all-around (GAA) nanowire. It is shown that an optimum cross section exists due to a trade-off between electrostatic and confinement. Also, the fundamental limit of improving gate control by thinning gate oxide is shown when passing from the equivalent oxide thickness (EOT) to the capacitive equivalent thickness (CET) concept. In Section 2.3, the physics and the performance limits of the new multibarrier boosted RT-FET are investigated through quantum simulations in silicon nanowires and compared to those of a nanowire multigate SOI MOSFET. Finally, the possibility of implementing RT-FET with ultra-low source-drain resistance dopant-segregated Schottky barrier MOSFETs is investigated in Section 2.4.

2.1   SIMULATION ALGORITHM

For ultra-scaled devices with cross section dimensions smaller than 10 nm and a gate length below a few tens of nanometers, quantum effects are playing a crucial role in device performances and parameters. Hence, the need for quantum simulations arises. Among the new simulation methods developed for that purpose, the nonequilibrium Green's function (NEGF) method [9–14] has gained popularity and shows a real potential for modeling quantum effects at the scale of a few nanometers. Computations that use this method can, however, be time-consuming, which is the main obstacle to its use in intensive device simulations. In an attempt to reduce simulation time, mode-space (MS) methods, which can result in a simulation speed-up up to two to three orders of magnitude over more computationally demanding real-space methods, have been introduced. Such an approach is assumed in the following. We note, however, that here we do not imply from MS uncoupled mode space (that is only valid for devices with small film thicknesses and no discontinuities in their cross section along the channel; see below) as sometimes assumed in the literature, but that the coupled mode-space approach is also encompassed.

Our quantum simulator is based on the use of a fast coupled mode-space (FCMS) implementation of the NEGF [7], adaptive energy, and nonuniform mesh algorithms. Compared to a real-space NEGF algorithm, the coordinates y and z of the cross section perpendicular to the transport direction, x, are replaced by the mode energies $E_{sub}^m(x)$ of the electron subbands in an MS approach. This drastically reduces computation time, as in practice only the first few subbands are populated by electrons and need to be taken into account [12]. A full description of our simulator can be found in [7]. Here we summarize the main equations and simplifying assumptions used by our 3D MS-NEGF simulator [15]. The main convergence loop of the program self-consistently computes the electrostatic potential in the device, V_1, by solving for the Poisson equation and the electron concentration, n_1, using the NEGF formalism [11]. A nonlinear Poisson scheme is used to ensure fast convergence. Except for the Schottky barrier case, which is treated below, Neumann (close) boundary conditions are used for the Poisson equation at the source and drain. In quantum simulations, indeed, the applied bias is fixed through fixing the Fermi level at source E_{FS} and draining ($E_{FD} = E_{FS} - qV_D$). This allows for electrons and potential to self-consistently adjust for ensuring charge neutrality. The mode-space longitudinal coupled Schrödinger equation from which the mode-space device Hamiltonian, H_{MS}, can be computed is given by [12]

$$-\frac{\hbar^2}{2}\bar{a}_{mm}(x)\frac{\partial^2}{\partial x^2}\varphi^m(x) - \sum_n E_C^{mn}(x)\cdot\varphi^n(x) + E_{sub}^m(x)\cdot\varphi^m(x) = E\varphi^m(x) \quad (2.1)$$

where

$$\bar{a}_{mm}(x) = \int_{y,z} \frac{1}{m_x^*(y,z)} \left| \xi^m(y,z;x) \right|^2 dydz \tag{2.2}$$

is the inverse of the average value of the effective mass in the cross section. $E_C^{mn}(x)$ is a term that represents the coupling between the lateral modes m and n and depends on the integral of the product between the transversal wave function ξ^m and the first and second derivatives of ξ^n in the cross section but is independent of y and z [12]. In the MS method, the subband energy profile $E_{sub}^m(x)$ and the corresponding transversal wave function $\xi^m(y,z;x)$ need to be calculated. This is done by solving a 2D Schrödinger problem in the cross section of the device at each slice $x = x_0$:

$$H_{2D}\xi^n(y,z;x_0) = E_{sub}^n(x)\xi^n(y,z;x_0) \tag{2.3}$$

and

$$H_{2D} = -\frac{\partial}{\partial y}\left(\frac{\hbar^2}{2m_y^*(y,z)} \frac{\partial}{\partial y} \right) - \frac{\partial}{\partial z}\left(\frac{\hbar^2}{2m_z^*(y,z)} \frac{\partial}{\partial z} \right) + U(x_0,y,z) \tag{2.4}$$

where $U = E_{CB} - q.V_1$ is the potential energy and E_{CB} is the bulk material conduction band edge (E_{CB} of Si is our potential reference and has been set to 0). In nanowires with a small and constant cross section, it is usually assumed that the eigenfunctions $\xi^n(y,z;x)$ remain constant along the channel even though the eigenvalues $E_{sub}^n(x)$ do vary. In this case the coupling term in Equation (2.3) would disappear and one can use the fast uncoupled mode-space (FUMS) hypothesis to further fasten the algorithm by computing $\xi^n(y,z)$ and replacing $U(x,y,z)$ by its x-averaged value in Equation (2.6). [12]. However, here a more general but slower coupled mode-space (CMS) approach is wanted, as in RT-FETs or Schottky barrier (SB) devices, tunnel barriers create strong variation of the wave function shape in their vicinity and therefore mode coupling. However, this perturbation is very local, and by considering coupling only in the vicinity of the barrier in the FCMS, we therefore achieve FUMS speed with accuracy of the CMS [7].

From H_{MS}, the mode-space device Hamiltonian, the retarded Green's function, G, of the active device can be defined, and from there, electron concentration and current in the device, as well as their energy spectrum, can be calculated using the NEGF approach [11, 12]:

$$G(E) = [EI - H_{MS} - \Sigma_S(E) - \Sigma_1(E) - \Sigma_2(E)]^{-1} \tag{2.5}$$

where Σ_s is the self-energy that accounts for scattering inside the device (in the ballistic case, Σ_s is equal to zero), and $\Sigma_1(\Sigma_2)$ is the self-energy caused by the coupling between the device and the source (or drain) reservoir [11].

One of the strengths of the NEGF is its ability to handle different types of elastic or inelastic scattering as electron-phonon or surface roughness scattering without using an averaged relaxation time approximation, as has been common in semiclassical or quantum-corrected drift diffusion, in higher-order Boltzmann moment equation, or in Monte-Carlo simulation, for instance. In the NEGF, indeed, spectral functions like density of state $DoS(x,E)$ or density matrixes $\rho(x,x,E)$ are computed self-consistently over a discretized energy mesh. The scattering rate, or scattering self-energy $\Sigma_S(x,x,E)$ matrix that depends on these spectral quantities and influences them can therefore also be calculated self-consistently at each energy, and in turn related to the exact band structure and carrier's population. This is increasingly important as transistors are scaled down and confinement makes the latter to become a strong function of the exact device structure and bias voltages. In the NEGF approach, the input scattering parameters are therefore not relaxation time or mobility—these are derived parameters that can be obtained as a result after convergence of the self-consistent loop and energy averaging— but directly the perturbative Hamiltonian, e.g., the electron-phonon interaction Hamiltonian for phonon scattering. The degree of accuracy in the modeling of this Hamiltonian results in a trade-off between simulation time and accuracy. We have developed such an approach to include phonon scattering (both elastic and inelastic) based on the self-consistent Born approximation and deformation potential theory [13, 14].

Finally, the methodology used to simulate Schottky contacts is similar to [16]. The Schottky barrier is added as a boundary condition in the source and drain potential. A potential equal to

$$V_{SB} = -q*(SB_H - (E_{CB} - E_F)) \tag{2.6}$$

is added to the source and drain potentials, where E_F is the Fermi potential related to the doping in the Si body, and E_{CB} and SB_H are respectively the bottom of the conduction band and the Schottky barrier height value in bulk silicon. This allows one to take into account the increase of the SB_H due to the increase of E_C through quantum confinement, which has been shown to be the dominant change in small cross sections. Note, however, that this is a worst-case scenario, as other effects should slightly counteract this increase of SB_H [17]. We have neglected these effects, however, due to the lack of experimental studies on the SB_H values in nanoscale Si devices and the fact that the trends should not change fundamentally with a change of a few tens of meV, as we have observed when comparing barriers differing by hundreds of meV [6, 8]. The conduction band edge in the Schottky metal, E_{CSB}, is assumed to be constant and a few 100 meV lower than the source and drain Fermi level in order to ensure sufficient injection (and independent of the exact band edge level of the metal) of carriers in the device.

2.2 GATE COUPLING OPTIMIZATION IN NANOSCALE NANOWIRE MOSFETs: ELECTROSTATIC vs. QUANTUM CONFINEMENT

Using our simulation tools, we now investigate the effects of electrostatics and quantum confinement in order to find an optimal cross section size for 10 nm channel length nanowires. Figure 2.1 shows the maximum film thickness that can be used vs.

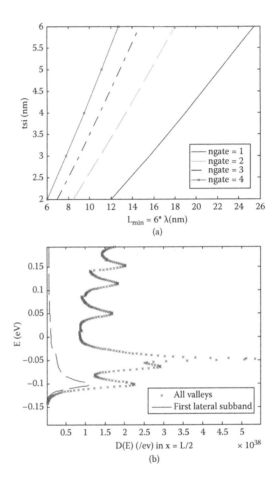

FIGURE 2.1 (a) Maximum film thickness vs. minimal channel length for one to four gate architectures following classical electrostatic theories and for an equivalent gate oxide thickness (EOT) of 0.5 nm. It is based on the natural length λ (Equation (2.7)) that scales down with the number of gates n_g. A minimum channel length of at least 6*λ must be taken to ensure good gate control and low short channel effects. (b) DoS vs. energy for a 4 × 4 nm^2 [100] GAA nanowire. The peak structure is related to the splitting of the conduction band in subbands due to 1D confinement.

the channel length for one to four gate architectures following classical electrostatic theories [18] and for an equivalent gate oxide thickness (EOT) of 0.5 nm. It is based on the natural length λ that scales down with the number of gates n_g:

$$\lambda = \sqrt{\frac{\varepsilon_{si}}{n_g \varepsilon_{ox}}\left(1 + \frac{\varepsilon_{ox} t_{si}}{\varepsilon_{si} t_{ox}}\right) t_{si} t_{ox}} = \sqrt{\frac{\varepsilon_{si}}{n_g \varepsilon_{SiO_2}}\left(1 + \frac{\varepsilon_{SiO_2} t_{si}}{\varepsilon_{si} \cdot EOT}\right) t_{si} \cdot EOT} \quad EOT = \frac{\varepsilon_{SiO_2}}{\varepsilon_{ox}} t_{ox} \quad (2.7)$$

For given values of Si and oxide thicknesses (t_{si} and t_{ox}, respectively) and permittivities (ε_{si} and ε_{ox}), a minimum channel length of at least 6*λ must be taken to ensure good gate control and low short channel effects. As can be seen for $L = 10$ nm a single gate device would require an EOT below 0.5 nm or a Si film thickness below 2 nm. Both options must, however, be discarded due to quantum effects.

Due to tunneling, a physical gate oxide thickness t_{ox} below 1.5–1 nm must be excluded to avoid high leakage due to gate tunneling currents. As gate control does not depend on t_{ox} alone but on t_{ox}/ε_{ox}, a high-k dielectric with a significantly higher permittivity than SiO_2 can achieve an equivalent oxide thickness (EOT) lower than this t_{ox} value. However, due to quantum confinement in film thicknesses of a few nm, a dark space region (the electron channel is not directly at the Si-oxide interface but at a depth t_d in the Si film) and a reduction of the DoS are observed [19]. Both effects tend to reduce the intrinsic gate capacitance and therefore the gate coupling. In order to take this effect into account, one can replace EOT by an equivalent capacitive thickness CET in Equation (2.7). The point is that CET tends to saturate when EOT is very small because dark space and DoS reduction does not scale down with EOT, so that even if EOT tends to 0, a minimum value of CET is reached [19]:

$$CET_{min} = CET(EOT \rightarrow 0) = \frac{t_d}{\lambda} \cdot \frac{\varepsilon_{Sio_2}}{\varepsilon_{Si}} \approx 0.55 \text{ nm} \qquad (2.8)$$

Because of quantum confinement ($t_d > 0$) and the low 1D DoS ($\lambda < 1$), it is not possible to have a better electrostatic control of the gate to the channel in nanowires of small cross sections than that achieved with an SiO_2 oxide of thickness CET_{min}, even if one could use an "ideal" gate dielectric with infinite permittivity. The value $CET_{min} = 0.55$ nm is obtained for Si by assuming $t_d = 0.8$ nm and $\lambda = 0.5$ at high V_g, which compares well with our simulation results for cross sections between 2×2 and 4×4 nm^2 and EOT smaller than 1 nm [19].

Concerning t_{si}, it is usually accepted that below 2 nm of film thickness the dependency of bandgap, and therefore threshold voltage, with t_{si} related to quantum confinement is too strong and would lead to too much variability [20]. The smaller the cross section, the larger the bandgap mismatch is for a given diameter variation, which would arise, for example, from process variability. This can be observed in Figure 2.2, which shows bandgap increase extracted from our 3D-NEGF simulations vs. the diameter reduction Δt_{si}. It can be quite well explained by using a simple analytical model of the energy level of a constant potential well with infinite potential barrier at the Si/SiO$_2$ interface. Using the effective mass approximation (parabolic E-k dispersion relationship) in a device with film thickness t_{si} and width 2_{si}, E_1, the first level above the conduction band. E_C, is then given by

$$E_1^{parab}(t_{si}, W_{si}) - E_C = \frac{\hbar^2}{2}\left(\left(\frac{\pi}{m_y^* t_{si}}\right)^2 + \left(\frac{\pi}{m_z^* w_{si}}\right)^2\right) \qquad (2.9)$$

When the diameter of the cross section is smaller than 3 nm, however, the E-k dispersion relationship is no longer parabolic and a correction factor has to be taken into account in order to obtain accurate results [21], but the trend of very high bandgap sensitivity to the exact dimension in small cross section given by Equation (2.9)

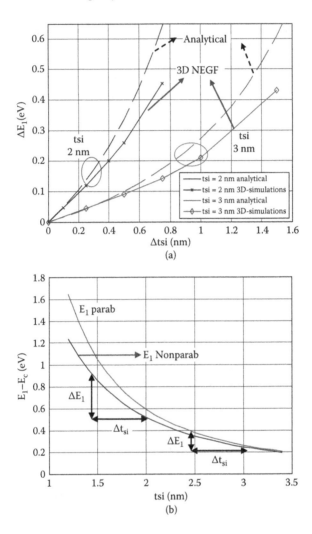

FIGURE 2.2 (a) Bandgap increase vs. diameter reduction of the cross section (Δt_{si}) for a reference diameter of 2 and 3 nm in a square nanowire. The diameter of the cross section after reduction is $t_{si} - \Delta t_{si}$, while its area is equal to $(t_{si} - \Delta t_{si})^2$. (b) Analytical calculation of the first energy level in a 2D infinite barrier potential well vs. cross section diameter with and without (Equation (2.16)) nonparabolic correction.

remains valid (Figure 2.5b). Figure 2.5a also shows a comparison of the simulated and analytically predicted bandgap variation assuming the correction factor given in [21] is used in both the simulations and the analytical model. The agreement is quite good and the difference can be explained by the penetration of the electron wave function in the oxide in the 3D-NEGF case.

Therefore, following the rule shown in Figure 2.1a, we see that a multigate architecture would be needed for scaling channel length down to 10 nm and below. For a square GAA (four gates) device, the cross section characteristic dimension should be below 5 nm for $L = 10$ nm. This is, however, based on a classical model that does not take into account quantum effects, while for such small cross section size values, effects like quantum confinement are quite important, as we have just seen above. Nevertheless, when looking at the characteristics of the simulated GAA nanowires of Figure 2.3a, we can see that in accordance to the natural length model, the 6×6

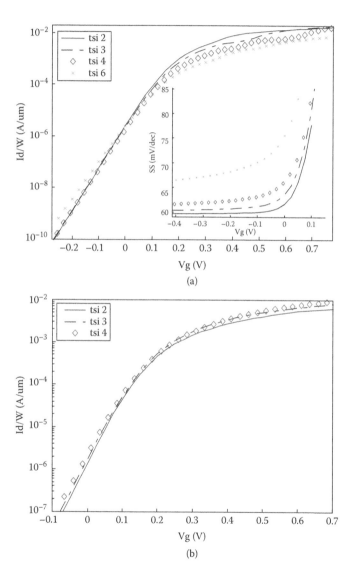

(a)

(b)

FIGURE 2.3 3D NEGF simulated drain current I_D vs. V_G of square [100] GAA nanowire MOSFET transistors for different film thickness t_{si} (nm). (a) Ballistic simulations. (b) Elastic and inelastic electron-phonon scattering simulations within self-consistent Born approximation. $L = 10$ nm, $V_D = 0.7$ V, $V_{th} = 0.19$ V. For case (a) subthreshold slope $SS(V_G)$ is also shown (inset).

nanowire has its subthreshold and on-current significantly degraded compared to the smaller cross section devices due to insufficient gate control. However, based on a pure electrostatic model, a smaller thickness would always improve electrostatic control, and hence I_{ON}/I_{OFF} ratios, while as we can see from the simulation results of Figure 2.3, this is not the case when taking quantum effects into account.

When looking at ballistic simulation first (Figure 2.3a), the improvement of sub-threshold slope (SS) with smaller t_{si} is indeed observed. Above threshold, however, we can see that the current does not increase continuously but presents some satura-tion steps. This is related to the subband structure caused by the lateral confinement and the resulting 1D-DoS (Figure 2.1b) that can create oscillation in current and capacitance of the nanowires [19]. When V_g is increased above threshold, the first subband eventually becomes full, and any further increase in V_g no longer results in an increase of the electron concentration: the $DoS(E)$ now decreases and the electron concentration saturates, which yields the saturation steps in the I_D-V_G of Figure 2.3a. As the gate voltage is further increased, however, higher-energy subbands start to become populated and the current increases again as higher-order subbands increase the electron concentration. A smaller cross section size allows for a better electro-static control, and therefore a faster filling with V_G of the first subband. We can see that the current increases first faster with V_G for smaller t_{si}. However, a smaller cross section size also implies stronger confinement, and therefore increased distance between the different subbands and longer saturation steps. This is why the current of the 3 × 3 and 4 × 4 nm^2 nanowires can increase faster and get comparable (or even slightly higher in the 3 × 3 nm^2 case) results to those of the 2 × 2 nm^2 nanowire at sufficiently high gate voltage. In the case of the 6 × 6 nm^2 nanowire, the electrostatic is really too bad and the on-current stays significantly lower.

When considering electron-phonon elastic and inelastic scattering using a self-consistent Born approximation, we can see a further effect of the confinement. In a 10 nm long nanowire, scattering does not strongly affect subthreshold characteristics, while intervalley inelastic scattering can degrade the on-current significantly, especially for ultra-thin cross section (below 4 × 4 nm^2). Due to the increased confinement, the elec-tron-phonon wave function overlap significantly increases, which in turn increases the scattering rate. Experimentally, mobility has been shown to drop rapidly when reduc-ing nanowire cross section below 4 × 4 nm^2 [22]. Figure 2.3b indeed shows that when considering scattering, the current in the 3 × 3 nm^2 cross section is now higher than in the 2 × 2 nm^2 cross section nanowire for every V_G above threshold. Note also that as spectral quantities like DoS are broadened due to scattering, the effect of the succes-sive subband filling is smoothed out and no longer visible in the $I_D(V_G)$ characteristics of Figure 2.3b.

Finally, when taking into consideration both electrostatics and quantum confine-ment, a cross section on the order of 3 × 3 nm^2 gives the best I_{ON}/I_{OFF} ratios.

2.3 PHYSICS OF RTFET

Despite the fact that we can get nearly ideal gate coupling in a well-scaled nanowire, the current increase rate with V_G, i.e., the slope of the $I_D(V_G)$ characteristics, is still limited by the way carriers are made available to drive a current when increasing V_G.

This depends on carrier statistics over energy, and because in a standard transistor the density of state close to the top of the channel barrier is about constant with V_G, when increasing (in an N-MOSFET) the gate voltage the current therefore increases at a rate dictated at best by Fermi-Dirac statistics only [5]. In order to further increase I_{ON}/I_{OFF} ratios, we have recently proposed using quantum effects to render this DoS nonconstant with V_G using a multibarrier gate-controlled transistor, the gate-modulated resonant tunneling FET (RT-FET). It is a MOSFET boosted with additional tunnel barriers (i.e., barriers of a few nanometers width and less than 10 nm) near the gate edge(s) and under electrostatic control of the gate that creates additional longitudinal confinement in the device [5]. Such TBs can be created, for instance, in a planar technology using quantum confinement and a local reduction, or constriction, of the device cross section (Figures 2.2 and 2.4a), resulting from a local oxidation that can be well controlled [7].

Figure 2.4a shows a schematic device representation and gives the parameters of the SOI gate-all-around n-channel nanowire with constrictions of width L_C and section reduction Δt_{si}. Using constrictions, barriers between a few meV and several hundred meV can be obtained by tuning Δt_{si} (Figure 2.2). An overlap covering the constrictions is required in order to keep adequate electrostatic control of the gate over the tunnel barriers (TBs) [5].

As observed in Figure 2.5, owing to the additional barriers and the related longitudinal confinement, the density of state (DoS) in an RT-FET is reduced in its off-state, while remaining comparable in its on-state to that of a MOS transistor without barriers. The RT-FET thus features both a lower RT-limited off-current and a faster increase of the current with V_G, i.e., an improved slope characteristic, and hence an improved I_{ON}/I_{OFF} ratio. The DoS being a function of position, the equivalent and more rigorous concept of transmission T(E) will be used for the current; i.e., the source-injected current spectrum J(E) (we neglect the drain injected current here for simplicity, assuming V_D greater than a few kT/q) is proportional to $T(E) \cdot f_{FD}(E - E_{FS})$. In a well-optimized RT-FET, in a subthreshold regime (Figure 2.5a), the channel and tunnel barriers being above E_{FS}, the high nonconfined transmission states above the well are filtered by the Fermi-Dirac statistics. The current is therefore flowing through the few first quasi-bound or resonant tunneling states in the well and becomes very low compared to a standard MOSFET.

When increasing V_G, however, the channel and tunnel barriers are pushed down in energy. When the TBs pass below E_{FS}, the filtering action of the Fermi-Dirac statistics vanish and high nonconfined transmission states above the well start to drive a significant amount of thermionic current (Figure 2.2b). This gives RT-FETs two different thresholds. The first, V_{th1} (= 0.19 V in Figure 2.4b), related to the resonant tunneling (RT) current, happens like in a standard transistor when the top of the channel barrier (TCB) passes below the source Fermi level E_{FS}. The second, V_{th2} (= 0.43 V in Figure 2.4b), related to the thermionic current above the well, happens when the TB passes below E_{FS}. For $V_G \geq V_{th2}$, an important additional thermionic current will start flowing, enabling further improvement of the slope and current ratio with V_G, and hence very high on-current (Figure 2.4b and Figure 2.5b). The transistor is recovering the on-current level of a MOSFET.

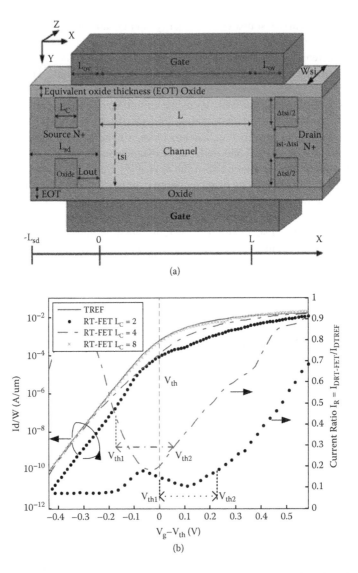

(a)

(b)

FIGURE 2.4 (a) The channel is assumed along the [100] crystallographic orientation. Gate-all-around SOI N nanowire with constrictions (the lateral gates are not shown). $t_{si} = w_{si} = 2$ nm. Channel: $L = 10$ nm, doping $N^- = 10^{15}$ cm^{-2}. S/D extensions: $L_{sd} = L_{ov} + 7$ nm, doping N$^+ = 10^{20}$ cm^{-3}. Oxide: EOT = 0.5 nm. $T = 300$ K. (b) Drain current $I_D(V_G)$ vs. V_G-V_{th} of TREF (reference nanowire MOSFET transistor, i.e., without constrictions) (1) and RT-FETs with $L_C = L_{ov} = 2$ nm (2), $L_C = L_{ov} = 4$ nm (3), and $L_C = L_{ov} = 8$ nm (4). $V_d = 1$ V, $L = 10$ nm. The current ratio, $I_R(V_G)$, is also shown for cases (2) and (3). The threshold voltage, V_{th}, of each device is extracted by considering the maximum of the second derivative of its $I_D(V_G)$ curve. V_{th} is in between V_{th1}, the channel barrier-related threshold voltage, and V_{th2}, the tunnel barrier (TB)-related threshold voltage. TB_S = 0.2 eV, TB_D = 0.41 eV, L_{out} = 1 nm. (From A. Afzalian J.-P. Colinge, and D. Flandre, Physics of gate modulated resonant tunneling (RT)-FETs: Multi-barrier MOSFET for steep slope and high current. *Solid-State Electronics* 59(1), 2011, pp. 50–61, doi: 10.1016/j.sse.2011.01.016.)

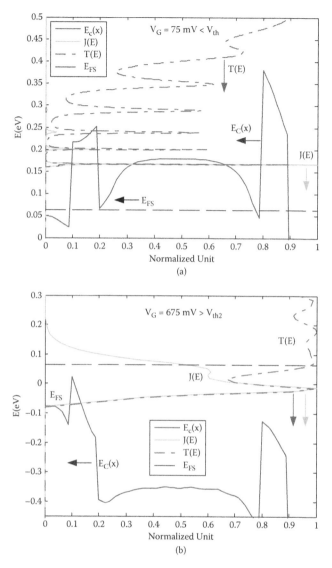

FIGURE 2.5 Conduction band profile E_C vs. normalized distance $((x - L_{sd})/(L + 2*L_{sd}))$, normalized current density spectrum J(E), and transmission spectrum T (nonnormalized) vs. energy (i.e., T and J curves are rotated by 90°) for RT-FET with $L_{out} = 1$ and $L_{ov} = 2$ nm. The position of the source Fermi level, E_{FS}, is also shown for comparison. (a) Below threshold: The nonconfined transmission states above the well are filtered by the Fermi-Dirac statistics. The current is therefore flowing through the few first quasi-bound or resonant tunneling states in the well and becomes very low compared to a standard MOSFET. (b) Above V_{th2}, when the tunnel barriers pass below E_{FS}, the filtering action of the Fermi-Dirac statistics vanishes and nonconfined transmission states above the well start to drive a significant amount of thermionic current. (From A. Afzalian J.-P. Colinge, and D. Flandre, Physics of gate modulated resonant tunneling (RT)-FETs: Multi-barrier MOSFET for steep slope and high current. *Solid-State Electronics* 59(1), 2011, pp. 50–61, doi: 10.1016/j.sse.2011.01.016.)

The RT-limited subthreshold regime also drives other interesting specifics characteristics: RT-FETs are intrinsically immune to source-drain tunneling; i.e., diffuse tunneling under the channel barrier that significantly increases off-current in a standard MOSFET with channel length below 10 nm is quantum mechanically excluded (Figure 2.6). RT-FETs therefore appear to be a promising candidate for

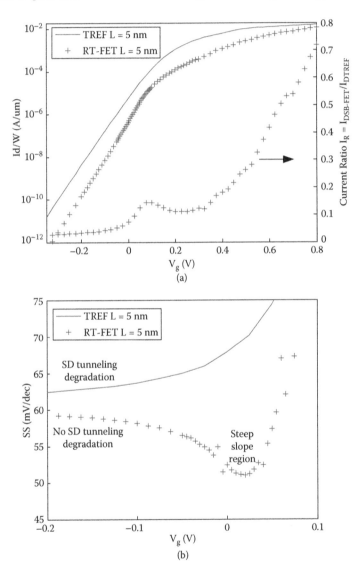

FIGURE 2.6 $I_D(V_G)$ and $SS(V_G)$ curves of TREF, and RT-FET with $L_{out} = 2$ nm and $L_{ov} = 3$ nm. $TB_S = 0.2$ eV, $TB_D = 0.41$ eV, $V_D = 0.7$ V, $L = 5$ nm, $L_C = 2$ nm, $t_{si} = w_{si} = 2$ nm. The current ratio, $I_R(V_G)$, is also shown and further enlightens the interesting properties of RT-FETs, i.e., low leakage, steep slope, sharp turn-on, and high on-current levels. Also, contrary to the MOSFET, no degradation of the subthreshold slope due to SD tunneling is observed in RT-FETs.

extending the roadmap below 10 nm channel length and boosting the on-current with alternative lower effective mass channel materials, as lower effective mass also results in increased SD tunneling. Subthreshold slope below the kT/q limits is possible through gate modulation of the RT states and the resonance condition, allowing for new resonant levels to drive the current when increasing V_G (Figure 2.6b). Under certain conditions it is also possible to create a zone of negative resistance, which is of interest for memory application, for example [5, 6].

2.3.1 INFLUENCE OF BARRIER WIDTH

In Figure 2.4b, $I_D(V_G)$ curves of RT-FETs with tunnel barriers, with increasing L_C and overlap L_{ov}, are shown and compared to those of an identical classical "reference" nanowire MOSFET without barrier or overlap (noted TREF below). We also show the current ratio of the two devices $I_R = I_{DRTFET}/I_{DTREF}$ vs. V_G. Depending on the tunnel barrier's width, L_C, the subthreshold current can be carried by resonant tunneling states (corresponding to sharp peaks in the transmission) in the well (RT current) or free or quasi-free states above the well (thermionic-like current) (both currents flow in case of RT-FET of Figure 2.7a). A transistor already dominated by the thermionic current below threshold, i.e., with L_C too large, will have characteristics very similar to a MOSFET but with a shifted threshold voltage, i.e., $V_{th} = V_{th2}$ (e.g., transistor with $L_C = 8$ nm in Figure 2.7b). A transistor dominated by the RT current below its threshold voltage, V_{th}, can achieve low off-current and steep slope region owing to the RT effect. Its actual V_{th} will be equal to V_{th1}. The RT-FET with $L_C = 2$ nm has the lowest I_{OFF} in the whole V_G range. However, for $V_G \geq V_{th2}$, an important additional thermionic current will start flowing (e.g., transistor with $L_C = 2$ nm) (Figure 2.4b), ensuring enhanced I_{ON}/I_{OFF} ratio with a good delay characteristic compared to the reference MOSFET. Transistors having intermediate L_C (i.e., $L_C = 4$ nm in Figure 2.4b) will have both thermionic and RT current flowing in subthreshold (Figure 2.7a). They will present a regime between V_{th1} and V_{th2} where their slope and current characteristics are in between subthreshold and above threshold. Their effective threshold voltage, V_{th}, will also be in between V_{th1} and V_{th2}. Although their off-current compared to RT-FET with $L_C = 2$ nm is not as low, they feature very sharp turn-on just above threshold, which ensures the best delay characteristics and comparatively good I_{OFF} performances for very low threshold voltages (i.e., 0.1 V and below). This makes them the most suitable devices for ultra-low voltage, especially at ultra-low threshold for high-speed application where a lower I_{ON}/I_{OFF} ratio can be traded off for low delay and high I_{ON}. Note also that the capacitance, or charge variation, increase of the RT-FETs due to the overlaps is more or less compensated by a charge reduction owing to the DoS reduction in the TBs (there are very few electrons in the tunnel barriers) and the quantum well related to the RT effect.

2.4 SB RT-FET

In order to achieve good performance RT-FETs in Si, their tunnel barriers should have a length, L_C, of a few nanometers, which is presently quite challenging to process. Replacing highly doped source (S) and drain (D) by metallic Schottky contacts,

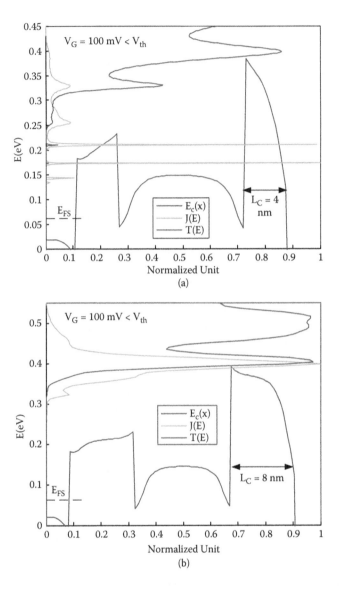

FIGURE 2.7 E_C vs. normalized distance, and J(E) (normalized) and T(E) (nonnormalized) vs. energy below threshold for RT-FET in subthreshold regime. (a) With $L_C = L_{ov} = 4$ nm. Due to the thicker L_C compared to $L_C = 2$ nm in Figure 2.2a, both thermionic and RT current are flowing in subthreshold. (b) With $L_C = L_{ov} = 8$ nm. Due to the even thicker L_C, only thermionic current is flowing in subthreshold and no on-regime steep slope region is observed in Figure 2.1b. (From A. Afzalian J.-P. Colinge, and D. Flandre, Physics of gate modulated resonant tunneling (RT)-FETs: Multi-barrier MOSFET for steep slope and high current. *Solid-State Electronics* 59(1), 2011, pp. 50–61, doi: 10.1016/j.sse.2011.01.016.)

Schottky barrier (SB) transistors have recently attracted a lot of attention because, doing so, one can potentially solve the increasing problem of S/D resistance and dopant diffusion [3, 23–25]. This will become crucial for future ultra-scaled transistors [3]. Due to the energy difference—when compared to vacuum level—between the conduction band in the silicon and the Fermi level in the Schottky metal (i.e., the Schottky barrier height, SB_H), however, a potential shift appears in the conduction band at the interface between Si and Schottky metal after Fermi level alignment. This creates a barrier in the conduction path that degrades subthreshold slope and current. In order to improve performances, one can find a material with lower barrier height, such as Er for n-FETs or Pt for p-FETs [16, 24]. However, midgap materials like nickel silicides are presently easier to integrate with Si [25]. Using dopant segregation (DS) of dopants implanted in the Schottky metal, i.e., the natural migration and accumulation in the first Si few nanometers near the interface with the Schottky metal, one can reduce the equivalent barrier height significantly [23, 25]. In fact, as observed on our simulation results (Figure 2.8), it is more the width of the barrier that decreases as a depletion region is thinned when increasing the doping [6, 8].

Using DS, one can therefore control barrier width by adjusting the doping level in the DS region. Our simulations show that a window of parameters where thin barriers and adequate resonant tunneling effect should result from using Schottky contact and dopant segregation in a nanoscale device allow us to filter the off- more efficiently than the on-current, and therefore pave the way for steep slope, low S/D resistance electronics.

Such an optimization is shown for a double gate (DG) device with $L = 12$ nm and $t_{si} = 2$ nm in Figure 2.9. The body is intrinsic (except for the DS regions near the gate edge of length L_{DS} for the SB-DS transistors). For DS, a doped region with L_{DS} of a few nanometers and doping concentration on the order of a few 10^{20} cm^3 can be achieved [23, 25]. The involved mechanisms, although a bit more complicated due to the parabolic shape and nonconstant width and height with V_G in the SB case, are very similar to those observed in rectangular constriction-induced tunnel barrier RT-FETs when varying barrier width and an optimal doping exists (on the order of 3×10^{20} cm^{-3} in the case of Figure 2.9) corresponding to an optimal barrier width (not too thick and not too thin). The major difference in SB-RT-FET, when compared to a RT-FET with constrictions, is that in the SB case the barrier height at the interface is fixed (Figure 2.8). When increasing V_G, the SB barrier is thinned, but its height at the interface cannot be reduced, and therefore in SB transistors, the tunnel barriers never pass below E_{FS} as the height at the interface is fixed. This phenomenon was responsible for the second threshold and the recovering of the on-current level of a MOSFET in the constricted RT-FETs. A similar, but reduced, effect can still be observed in a SB-RT-FET (Figure 2.9). By increasing V_G the barriers are thinned. The thinner the barriers at a given energy, the higher the tunneling probability of electrons through the barriers, the higher the coupling to the contact, and the broader the peaks in the quantum well at this energy. One passes from a close system with very sharp peaks to an open system with continuous longitudinal (x) energy spectra for electrons, transmission, and current above this energy. For energies where the barrier is very thin, the longitudinal confinement is so weak that the characteristics are very close to those of TREF. In order to quantitatively take this

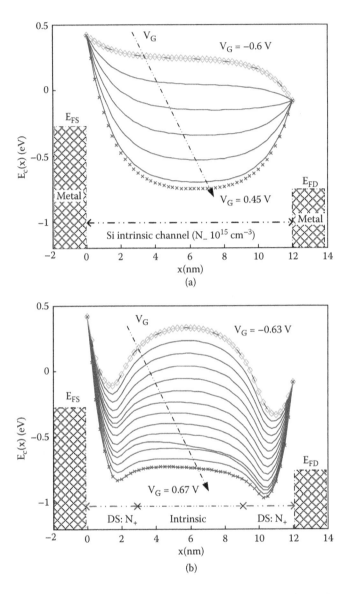

FIGURE 2.8 $E_C(x)$ vs. V_G for the SB-DG FET with $SB_H = 0.6$ eV (Ni-Si) (a) without DS and (b) with DS. States filled with electrons in the conduction band of the Schottky contacts, i.e., below E_{FS} and E_{FD}, are also shown.

phenomenon into account and therefore fairly compare MOSFETs and SB-RT-FETs, one must include scattering. This is done in the case of square cross section GAA nanowires in Figure 2.10.

Electron-phonon scattering does not strongly affect subthreshold characteristics of TREF or SB-FETs, while it affects the on-current. This is because the probability of scattering events increases with electron concentration. TREF current is thermionic

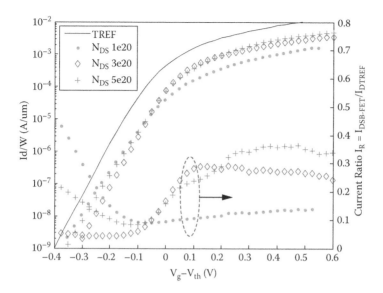

FIGURE 2.9 $I_D(V_G - V_{th})$ curves of the DG transistors without (TREF) and with Schottky contacts, with $SB_H = 0.28$ eV (Er-Si) and As dopant DS N_{DS} ranging from 1.10^{20} to 5.10^{20} cm^{-3}. $L_{DS} = 2$ nm, $Vd = 0.5$ V, $L = 12$ nm, $EOT = 0.5$ nm, $t_{si} = 2$ nm. The ratio SB-FET current/ TREF current, $I_R(V_G)$, is also shown. Gate work function shift between SB-FETs and TREF: $W_{SB\text{-}FETs} = 0.26$ eV ($= E_{FS,TREF} - E_{FS,SB\text{-}FET}$). $V_{th} = 50$ mV for all devices, except for SB-FET with $N_{DS} = 1.10^{20}$ cm^{-3}, where $V_{th} = 130$ mV.

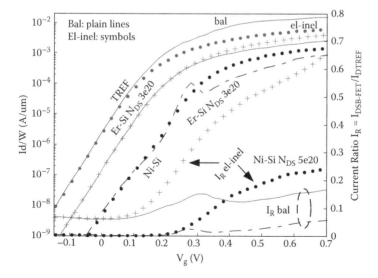

FIGURE 2.10 Comparison of $I_D(V_G)$ curves in the ballistic (bal, plain line) regime, and in the presence of elastic and inelastic (el.-inel., symbols) scattering of GAA nanowire without (TREF) and with Schottky contacts with (1) $SB_H = 0.28$ eV, $L_{DS} = 2$ nm, and $N_{DS} = 3.10^{20}$ cm^{-3}; (2) $SB_H = 0.6$ eV, $N_{DS} = 5.10^{20}$ cm^{-3}, and $L_{DS} = 3$ nm; and (3) $L = 12$ nm. EOT = 0.5 nm, $t_{si} = 2$ nm.

and scattering can only decrease it compared to the ballistic case. Therefore, its on-current is decreased by scattering. For the SB-FETs, the on-current is dominated by the resonant tunneling current peaks that are broadened as the barriers are thinned when increasing V_G. When scattering is considered, simulations results show that resonant peaks in the well are further broadened due to decoherence. The tunneling probability therefore increases faster with V_G compared to the ballistic case, hence the increased on-current. This allows the SB-FETs to have further improved I_{ON}/I_{OFF} ratios with comparatively good I_{ON} performances, especially in the Er-Si with $N_{DS} = 3 \times 10^{20}$ cm^{-3} case. The broadening effect of scattering can, however, also smooth out or make disappearing characteristics like steep subthreshold slope or negative resistance region. In the ballistic Ni-Si case, a steep subthreshold slope region on the order of 35 mV/dec, followed by a small region of NR, is observed (Figure 2.10), while, when considering scattering, the steep slope region is not as steep and the NR region is lost.

2.5 CONCLUSIONS

As transistors are scaled down in the nanoscale regime, scaling alone is not sufficient to achieve performance improvement and new boosters and device concepts are needed. One of the big challenges of scaling is the trade-off between power and performance that sets requirements on I_{ON} and I_{OFF} with a limited supply voltage. Therefore, optimizing the I_{ON}/I_{OFF} ratio and slope characteristics through device and device architecture optimization is utterly important. At this scale, however, quantum effects are playing a crucial role in device performances and parameters and must be taken into account when doing such an optimization. NEGF quantum simulation tools therefore appear as best suited to explore the physics and perform the optimization of nanoscale devices. We have reviewed here efficient NEGF algorithm methods suitable for intensive device simulations and used them to explore and optimize characteristics of ultra-scaled nanodevices. I_{ON}/I_{OFF} ratios and slope characteristics of transistors depend on the gate-to-channel coupling and carrier statistics that dictate the way carriers are made available to drive a current when increasing V_G.

We have first investigated the gate-to-channel coupling in a 10 nm channel length device. To ensure sufficient electrostatic control, ultra-thin-film nanowire and multigate architectures are the best suited. Quantum effects and their impact on ultra-scaled nanowires have been enlightened when optimizing I_{ON}/I_{OFF} ratios vs. the cross section size in a 10 nm gate-all-around (GAA) nanowire. It is shown that an optimum cross section of about 3×3 nm^2 exists due to a trade-off between electrostatic and confinement. Also, the fundamental limit of improving gate control by thinning gate oxide due to dark space and low 1D-DoS that do not scale with EOT has been shown and led us to switch from EOT to the CET concept.

As the density of state close to the top of the channel barrier is about constant with V_G in a standard transistor, when varying the gate voltage the current varies at a rate dictated by Fermi-Dirac statistics only. In order to further increase I_{ON}/I_{OFF} ratios, we have proposed a new device concept, the gate-modulated resonant tunneling FET (RT-FET), that uses quantum effects to render this DoS nonconstant with V_G

using a multibarrier gate-controlled architecture. The physics and the performance limits of this new multibarrier boosted RT-FET were investigated through quantum simulations in silicon nanowires and compared to those of a nanowire multigate SOI MOSFET. We have shown the possibility for a 10-nm-long RT-FET to improve the I_{ON}/I_{OFF} ratio by an order of magnitude, while keeping a similar on-current level, and therefore similar delays, as in a MOSFET. When scaling down channel length below 10 nm, the gain in I_{ON}/I_{OFF} ratios will further increase as RT-FETs are immune to SD tunneling. Finally, the possibility of implementing RT-FET with ultra-low source-drain resistance dopant-segregated Schottky barrier MOSFETs was also investigated, paving a way toward steep slope, low SD resistance devices.

ACKNOWLEDGMENTS

The author gladly acknowledges Prof. J.-P. Colinge from Tyndall National Institute, Ireland, and Prof. D. Flandre from UCL, Belgium, for the useful discussions and their support related to this work and its funding. This material is based upon works supported by FRS-FNRS Belgium.

REFERENCES

1. S. Borkar, Design challenges of technology scaling, *IEEE Micro*, July–August 1999, p. 23.
2. L. Kish, End of Moore's law: thermal (noise) death of integration in micro and nano electronics, *Phys. Lett. A*, 305, 144, 2002.
3. http://www.itrs.net/.
4. Q. Rafhay, R. Clerc, G. Ghibaudo, and G. Pananakakis, Impact of source-to-drain tunnelling on the scalability of arbitrary oriented alternative channel material nMOSFETs, *Solid-State Electron.*, 52, 1474, 2008.
5. A. Afzalian, J.-P. Colinge, and D. Flandre, Physics of gate modulated resonant tunneling (RT)-FETs: multi-barrier MOSFET for steep slope and high on-current, *Solid-State Electron.*, 59(1) May 2011, 50–61, doi: 10.1016/j.see.2011.01.016
6. A. Afzalian and D. Flandre, Breaching the kT/Q limit with dopant segregated Schottky barrier resonant tunneling MOSFETs: a computational study, presented at Proceedings of ESSDERC 2010 Conference, Sevilla, Spain.
7. A. Afzalian et al., A new F(ast)-CMS NEGF algorithm for efficient 3D simulations of switching characteristics enhancement in constricted tunnel barrier silicon nanowire MuGFETs, *J. Comput. Electron.*, 8, 287, 2009.
8. A. Afzalian and D. Flandre, Computational study of dopant segregated nanoscale Schottky barrier MOSFETs for steep slope, low SD-resistance and high on-current gate-modulated resonant tunneling FETs, *Solid-State Electron.* 65–66, Nov–Dec, 2011, 123–129, doi: 10.1016/jsse.2011.06.017.
9. P. Keldysh, *Sov. Phys. JETP*, 20, 1018, 1965.
10. P. Kadanoff and G. Baym, Diagram technique for nonequilibrium processes, *Quantum Statistical Mechanics*, Benjamin, New York, 1962.
11. S. Datta, Nanoscale device modeling: the Green's function method, *Superlattices Microstruct.*, 28, 253, 2000.
12. J. Wang, E. Polizzi, and M. Lundstrom, A three-dimensional quantum simulation of silicon nanowire transistors with the effective-mass approximation, *J. Appl. Phys.*, 96, 2192, 2004.

13. R. Lake and S. Datta, Non-equilibrium Green's-function method applied to double-barrier resonant-tunneling diodes, *Phys. Rev. B*, 45, 6670 1992.

14. S. Jin, Y. J. Park, and H. S. Min, A three-dimensional simulation of quantum transport in silicon nanowire transistor in the presence of electron-phonon interactions, *J. Appl. Phys.*, 99, 123719 (2006).

15. A. Afzalian, Computationally efficient self-consistent Born approximation treatments of phonon scattering for Couple-Mode Space Non-Equilibrium Green's Functions, *J. Appl. Phys.*, 110, 094517 (2011).

16. J. Guo and M. S. Lundstrom, A computational study of thin-body, double-gate, Schottky barrier MOSFETs, *IEEE Trans. Electron Dev.*, 49, 1897, 2002.

17. C. Tivarus, J. P. Pelz, M. K. Hudait, and S. A. Ringel, Direct measurement of quantum confinement effects at metal to quantum-well nanocontacts, *Phys. Rev. Lett.*, 94, 206803, 2005.

18. C.-W. Lee et al., Device design guidelines for nano-scale MuGFETs, *Solid-State Electron.*, 51, 505, 2007.

19. A. Afzalian, C. W. Lee, N. Dehdashti-Akhavan, R. Yan, I. Ferain, and J. P. Colinge, Quantum confinement effects in capacitance behavior of multigate silicon nanowire MOSFETs, *IEEE Trans. Nanotechnol.*, DOI: 10.1109/TNANO.2009.2039800.

20. W.-Y. Lu and Y. Taur, On the scaling limit of ultrathin SOI MOSFETs, *IEEE Trans. Electron Dev.*, 53, 1137, 2006.

21. J. Wang, A. Rahman, A. Ghosh, G. Klimeck, and M. Lundstrom, On the validity of the parabolic effective mass approximation for the I-V calculation of silicon nanowire transistors, *IEEE Trans. Electron Dev.*, 52, 1589, 2005.

22. S. D. Suk et al., Investigation of nanowire size dependency on TSNWFET, in *IEDM 2007*, p. 891.

23. Q. T. Zhao, U. Breuer, E. Rije, St. Lenk, and S. Mantl, *Appl. Phys. Lett.*, 86, 062108, 2005.

24. G. Larrieu and E. Dubois, Integration of PtSi-based Schottky-barrier p-MOSFETs with a midgap tungsten gate, *IEEE Trans. Electron Dev.*, 52, 2720, 2005.

25. S. F. Feste, J. Knoch, D. Buca, Q. T. Zhao, U. Breuer, and S. Mantl, Formation of steep, low Schottky-barrier contacts by dopant segregation during nickel silicidation, *J. Appl. Phys.*, 107, 044510, 2010.

3 SiGe BiCMOS Technology and Devices

Edward Preisler and Marco Racanelli

CONTENTS

3.1 INTRODUCTION

Over the past decade SiGe BiCMOS has evolved into a dominant technology for the implementation of radio frequency (RF) circuits. By providing performance, power consumption, and noise advantages over standard complementary metal-oxide-semiconductor (CMOS) transistor technology while leveraging the same manufacturing infrastructure, SiGe BiCMOS technologies can offer a cost-effective solution for challenging RF and analog circuit applications. Today many cell phones, wireless local area network (WLAN) devices, GPS receivers, and digital TV tuners employ some SiGe BiCMOS circuitry for either RF receive or transmit functions because of these advantages. Recently, advanced-node RF CMOS has achieved performance levels that enable some of these applications to be realized in CMOS alone, but SiGe BiCMOS continues to provide advantages for many leading-edge products. These existing markets, as well as emerging applications in the use of SiGe for power amplifiers and millimeter-wave products, continue to drive SiGe technology development.

In this chapter, we will review SiGe BiCMOS technology and its most significant applications. First, we will provide a basic understanding of how SiGe devices achieve a performance advantage over traditional bipolar and CMOS devices. Next, we review historical application drivers for SiGe technology and project a roadmap of SiGe applications well into the future. Next, we discuss RF performance metrics for SiGe heterojunction bipolar transistor (HBT) devices, followed by a discussion of how the devices can be optimized to maximize these performance metrics.

Finally, we discuss some of the components built around SiGe devices that are part of modern SiGe BiCMOS technologies and make them suitable for advanced RF product design.

3.2 SiGe HBT DEVICE PHYSICS

SiGe heterojunction bipolar transistor (HBT) devices are bipolar junction transistors created using a thin epitaxial base incorporating roughly 8% to 30% atomic germanium content. These devices are fabricated alongside CMOS devices with the addition of four to seven masking layers relative to a core CMOS process. SiGe HBTs derive part of their performance benefits from heterojunction effects and part from their epitaxial base architecture. Heterojunction effects were first described in the 1950s by Kroemer (eventually earning him a Nobel Prize), and more recently summarized by the same author [1]. These effects arise from the combination of different materials (in this case a $Si_{1-x}Ge_x$ alloy and Si) to create a variation in bandgap throughout the device that can be manipulated to improve performance.

Two common techniques for using heterojunction effects to improve performance are depicted in Figure 3.1, where typical doping and germanium profiles are shown along with the resulting conduction band energy profile. The first technique (c.f. Figure 3.1(a)) uses a box-shaped Ge profile. This creates an offset in the conduction band energy level at the emitter-base junction (due to the lower bandgap of SiGe relative to Si), lowering the barrier for electron current flow into the base, and increasing the efficiency of electron injection into the base. The band offset in the valence band is relatively unchanged compared to a silicon homojunction, and thus holes in the base are injected back into the emitter at roughly the same rate as they would be without the germanium. The combination of greater electron injection efficiency without also increasing the efficiency of the back-injection of holes from

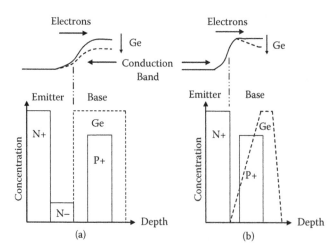

FIGURE 3.1 Common SiGe HBT doping and germanium profiles shown along with resulting band diagrams for (a) a box Ge profile and (b) a graded Ge profile.

the base results in higher current gain (collector current divided by base current, denoted as β for bipolar transistors). For a homojunction device, an increase in gain can be realized only by either thinning the metallurgical base width or increasing the doping in the emitter. The higher current gain in a SiGe HBT can then be traded off for increased base doping or lower emitter doping to improve base resistance and emitter-base capacitance, resulting in greater RF performance.

The second technique for utilizing heterojunction effects in an HBT (c.f. Figure 3.1(b)) employs a *graded* Ge profile to create a built-in (exists at zero bias) electric field in the base that accelerates electrons, reducing base transit time and improving high-frequency performance. This second technique somewhat off-sets the effects of the first technique [2] since using the graded profile necessarily means a reduction in the Ge content at the emitter-base junction, thus reducing the conduction band-lowering effect discussed above. Thus, careful design of the Ge profile throughout the device is a key factor in achieving optimal device perfor-mance. Today's SiGe bipolar HBTs make use of these two techniques to varying degrees to create a performance advantage over conventional bipolar devices.

Use of an epitaxially grown base rather than one formed by ion implantation is another reason SiGe HBTs exhibit better performance than conventional bipolar devices. The base of a conventional bipolar device is formed by implanting base dopant into silicon, which results in a relatively broad base after subsequent ther-mal processing. Epitaxy allows one to "grow in" the base doping profile through deposition of doped and undoped Si and SiGe layers controlled to nearly atomic dimensions. This allows the device designer to create an arbitrary base profile. An implanted device is limited to skewed Gaussian dopant profiles whose width is a function of implantation energy. Usually, the epitaxy technique is used to distribute the same base dose in a narrower base width, improving transit time through the base and resulting in better high-frequency performance.

Despite the advantages introduced by the epitaxial growth of the base layer, the final dopant profile in the device is largely determined by the subsequent thermal processing of the wafers after the base growth. Due to the large diffusion coefficient of boron (typically used as the base dopant) in silicon, a narrow as-grown base profile might be diffused dramatically by the time the processing is completed. Germanium itself actually serves to arrest the diffusion of boron somewhat, but in modern SiGe HBTs another atomic species, carbon, is added in the epitaxial base of SiGe devices to further arrest the diffusion of boron [3]. A small amount of carbon is added (typi-cally <1% atomic concentration of carbon is used) in the SiGe base during epitaxial deposition such that the electrical behavior is not significantly altered but the mate-rial properties are altered to reduce boron diffusion. The carbon helps to maintain a final boron profile closer to the as-deposited profile than it would be without the carbon. The electrical effect is a faster transit time due to a narrower base width, improving high-frequency performance. It should be noted that the introduction of carbon does reduce some of the beneficial band offsets introduced by Ge.

A final note about how the design of the epitaxial base growth affects HBT per-formance concerns strain. All SiGe HBTs are grown pseudomorphically on a silicon substrate, meaning that the SiGe (or in modern devices SiGe:C) is strained to take on the lattice constant of bulk silicon in the plane of the wafer. Any relaxation of the SiGe layers would generate dislocation-type defects that would short-circuit the emitter through the base to the collector of the device. Thus, all SiGe base layers must necessarily be grown pseudomorphically. Since bulk SiGe has a lattice constant larger than that of silicon, the SiGe is always under compressive strain in the plane of the wafer and the lattice stretches out in the direction perpendicular to the surface of the wafer according to the material's Poisson ratio. This strain can actually serve to enhance the mobility of both electrons traveling vertically through the device and holes traveling horizontally from the extrinsic base to the intrinsic base. In bulk $Si_{1-x}Ge_x$ the mobility of electrons is actually *lower* than that of bulk silicon until the germanium concentration gets close to 100%. However, strained SiGe can have electron mobilities equal to or even superior to those of bulk silicon [4], thus enhancing the transit of the minority carrier electrons through the base. The introduction of carbon mitigates some of this strain by pushing the SiGe:C layer's lattice constant back closer to bulk silicon. So again, trade-offs exist in the introduction of carbon in various locations of the epitaxial base growth.

In summary, modern SiGe HBT devices make use of the introduction of both Ge and C into the base of the transistor in order to manipulate both the electronic structure and metallurgical structure of the device to achieve performance not otherwise obtainable in bulk silicon devices.

3.3 APPLICATIONS DRIVING SiGe DEVELOPMENT

Several applications have driven advances in SiGe technology since the first high-speed SiGe bipolar devices were demonstrated in the late 1980s [5]. Initially, SiGe devices were conceived as a replacement for the Si bipolar device for emitter-coupled logic (ECL), high-speed digital integrated circuits (ICs) where SiGe transistors promised higher f_T, improving gate delay relative to their silicon bipolar or CMOS counterparts. However, advancements in the density, performance, and power consumption of CMOS technology quickly made it the logical choice for all but a few of these applications. So in the mid-1990s SiGe technology appeared to have a limited application base in only specialized very high-speed digital functions.

With the boom in wireless communications that began in the mid-1990s, however, a new application emerged as the primary driver for SiGe technology: the transceiver of a cellular phone. This application is tailor-made for SiGe BiCMOS. It requires good high-frequency performance to support carrier frequencies in the 900 MHz to 2.4 GHz range, very low-noise operation (as very small signals must be received and amplified), and a large dynamic range (as large output signals are required to drive the power amplifier and the antenna communicating with a far-away base station). In addition to wireless transceivers, high-speed fiber optic transceivers also provided a good application for SiGe transistors, as these pushed to even higher speeds, moving from 3 Gb/s to 10 Gb/s and eventually 40 Gb/s data rates. The transition from 3 Gb/s to 10 Gb/s provided a strong market for SiGe devices, as many of the same characteristics required for wireless transceivers are important in these transceivers (high speed, low noise, large dynamic range). But the transition from 10 Gb/s to higher rates was delayed with the dot-com bust, and 40 Gb/s and above are just now finally becoming mainstream technologies. Despite these higher data rate communication standards not becoming a reality, it was this expected transition in the late 1990s and early 2000s that pushed researchers to invest in creating very high-speed SiGe transistors (with f_T and f_{MAX} of 200 GHz and above) that are now poised to take advantage of perhaps other emerging applications.

Today, deep submicron CMOS is challenging SiGe for some of these traditional applications for two reasons: the speed of CMOS is adequate for many applications (although SiGe maintains an advantage in noise and an even wider advantage in dynamic range), and the density of CMOS is now high enough to enable new architectures that rely more heavily on digital signal processing rather than high-fidelity analog manipulation. In many cases, however, SiGe technology still offers a performance and power advantage and will continue to play a strong role in both the wireless and wire-line transceiver market. It should also be noted that, at present, the cost of advanced SiGe BiCMOS wafers is significantly less than that of the RF-performance-equivalent CMOS node since the SiGe BiCMOS devices don't rely on nearly as advanced lithography nodes as their CMOS counterparts. In addition to the incursion of CMOS devices in the traditional SiGe application space, current or even past-generation SiGe transistor performance is more than adequate to serve many of these applications. Therefore, these markets are becoming less important as drivers for future technology advancements.

Looking forward, however, two applications are primarily driving SiGe performance advancements today: higher-frequency millimiter-wave communications, and higher-power but lower-frequency products. Millimeter-wave applications include, for example, proposed ~60 GHz WLAN standards [6], 77 GHz automotive collision avoidance systems, 94 GHz and above terahertz passive imaging, and 40 to 100 Gb/s optical networking communications. These applications will serve to drive the speed of the SiGe transistor to higher and higher levels. At the other end of the performance vs. breakdown spectrum, high-power applications include, for example, the power amplifier for wireless devices and laser/optoelectronic modulator drivers for wire-line transceivers. These applications will drive improved trade-offs between speed and breakdown voltage in SiGe transistors. In the next section, we will review in more detail the design of SiGe transistors and see how improved speed and improved high-power performance are being realized.

3.4 SiGe PERFORMANCE METRICS

Two figures of merit are typically used to benchmark high-frequency device performance: (1) the cutoff frequency (f_T), which, for a bipolar device, is defined as the frequency at which the AC gain is unity, and (2) the maximum frequency of oscillation (f_{MAX}), which, for a bipolar device, is defined as the frequency at which the power gain is unity (usually the unilateral power gain).

For a bipolar device, f_T and f_{MAX} are related to basic device parameters by the commonly used equations

$$F_t = \frac{1}{2\pi \cdot \tau_F} \tag{3.1}$$

$$\tau_F = (C_{BC} + C_{BE}) \cdot \left(R_E + \frac{kT}{qI_C} \right) + \frac{W_B^2}{2D_B} + \frac{W_C}{2v_S} + R_C \cdot C_{BC} \tag{3.2}$$

$$F_{\max} = \sqrt{\frac{F_t}{8\pi \cdot R_B \cdot C_{BC}}} \tag{3.3}$$

where τ_F is the forward transit time, C_{BE} is the emitter-base capacitance, C_{BC} is the base-collector capacitance, R_E is the emitter series resistance, I_C is the collector current, W_B is the vertical base width, D_B is the electron diffusion length in the base, v_S is the electron saturation velocity, W_C is the vertical collector-base depletion width, and R_C is the collector resistance. At low collector current, f_T is dominated by the first term in Equation (3.2) where the junction capacitances combine with internal resistances to create an $R \times C$ time constant delay that is significantly longer than the other time constants in Equation (3.2) (see Figure 3.2). At high current the term W_B becomes a function of I_C. When the charge associated with the current through the collector-base depletion region becomes comparable to the intrinsic doping level on either side, the edges of the depletion region collapse, and thus "push" the depletion region away from the base, effectively widening the base. Mathematically, this occurs when

$$J_C \approx qN_C v_{SAT} \tag{3.4}$$

where N_C is the nominal doping in the collector and v_{SAT} is the electron saturation velocity in the collector. This base push-out, known as the Kirk effect [7], is responsible for f_T decreasing at high current rather than saturating, as would otherwise be predicted by Equation (3.2). So both f_T and f_{MAX} peak at a specific current density (see Figure 3.2).

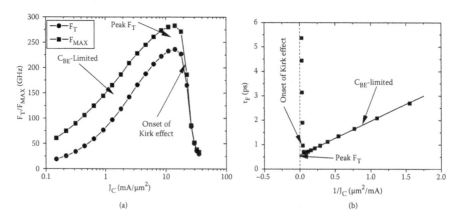

(a)   (b)

FIGURE 3.2 (a) Typical F_T vs. J_C plot for a state-of-the-art, volume-manufactured SiGe HBT showing the various regions indicated by the terms in Equation (3.2). (b) Typical τ_F ($1/(2\pi F_T)$) vs. $1/J_C$ plot indicating the various regions indicated by the terms in Equation (3.2).

Obtaining ever-higher peak f_T and f_{MAX} are important because, while today's volume RF applications target modest operating frequencies relative to the peak f_T values shown in Figure 3.2, high peak f_T (and f_{MAX}) can be traded off for other benefits, including reduced power consumption, higher breakdown voltage, and reduced noise.

Figure 3.3 shows an example of the power savings that can be achieved with higher f_T SiGe technology even when operating at relatively low frequencies. For instance, at 25 GHz or so there is a 3× improvement in current consumption going from a 0.3 μm technology to a 0.2 μm technology. At 50 GHz the advantage is 4×. Thus, the scaling of the emitter is a key factor in reducing the power consumption required for a given f_T. Alternatively, when used as a gain stage in an amplifier, one could operate the device at peak f_T and simply use fewer gain stages to achieve the same total circuit gain. For instance, in theory one could use any of the top three technologies shown in Figure 3.3 (0.13, 0.15, and 0.2 μm emitter width SiGe HBT technologies) to achieve gain at 100 GHz. However, the 0.13 μm technology provides approximately 7 dB of gain at 100 GHz for peak f_T conditions, whereas the 0.2 μm technology provides only 3.5 dB. Thus, you could reduce the number of gain stages by half to achieve the same total gain, which provides an advantage in terms of power consumption, circuit area, and the total noise added by the circuit.

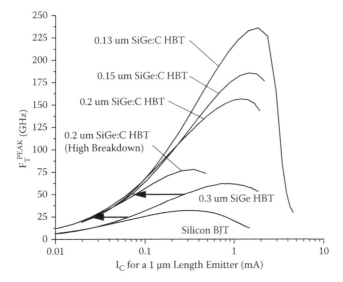

FIGURE 3.3 f_T for various TowerJazz BiCMOS technologies plotted as a function of I_c for a minimum width and unit length emitter. The dimensions in the labels refer to the minimum emitter width, not the corresponding CMOS technology node. In addition to higher peak f_T, subsequent technology nodes lower power consumption even when biasing at low f_T, as indicated by the arrows.

The second advantage of higher f_T values, even in lower frequency applications, is in the RF noise figure. A minimum noise figure can be expressed by [8]

$$NF_{min} = 1 + \frac{n}{\beta} + \frac{f}{F_t} \cdot \sqrt{\frac{2qI_C}{kT} \cdot (R_E + R_B) \cdot \left(1 + \frac{F_t^2}{\beta \cdot f^2}\right)} + \frac{n^2 F_t^2}{\beta \cdot f^2} \qquad (3.5)$$

where n is the collector current quality factor and β is the current gain. From Equation (3.5), it is seen that with a high β, as is typically seen in SiGe devices, the noise figure reduces to

$$NF_{MIN} \approx \frac{f}{F_t} \cdot \sqrt{\frac{2qI_C}{kT} \cdot (R_E + R_B)} \qquad (3.6)$$

In this limit a higher f_T and lower R_B result in a lower noise figure. Further, the ability to operate at a lower I_C and obtain the same f_T can lead to a reduction in the term under the radical in Equation (3.6). Thus, all of the advantages enabled by using SiGe in the base of an HBT are brought to bear when one is attempting to minimize a noise figure: the high β enabled by putting SiGe at the emitter-base junction, the lower R_B for a given β enabled by increasing the base doping, and the higher f_T enabled by the Ge profile in the base.

Finally, f_T can be traded off for higher breakdown voltage by modulating the collector doping concentration through a collector implant mask such that multiple devices spanning a range of f_T and breakdown are made available on the same wafer. Figure 3.4 shows the family of devices realized by this technique across several

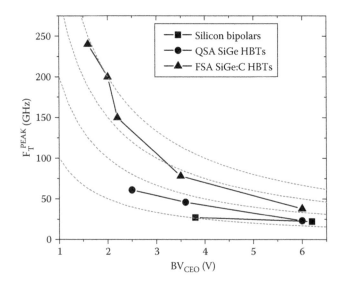

FIGURE 3.4 f_T vs. BV_{CEO} plotted for all devices available in several generations of bipolar technologies. The dashed lines are contours of constant $f_T \times BV_{CEO}$. All data are from TowerJazz electrical specifications.

generations of TowerJazz technology. Each subsequent generation supports devices with higher f_T but also improves the trade-off between f_T and breakdown voltage improving large signal performance for applications such as integrated drivers and power amplifiers. This is in contrast with CMOS, where each new generation makes the integration of power devices more difficult due to the more brittle gate oxide forcing lower voltage ratings.

3.5   DEVICE OPTIMIZATION AND ROADMAP

Higher f_T values are enabled by vertical scaling of the HBT device. The most fundamental device enhancement with each generation of higher f_T devices is scaling of the base width, the W_B term in Equation (3.2). The fundamental limit of scaling the base width occurs when the emitter-base and base-collector depletion regions touch and thus the device is "punched through." It should be noted that the metallurgical base width in advanced SiGe HBTs is already many times narrower than all but the most aggressive CMOS channel lengths. The next most commonly adjusted vertical scaling parameter is the collector doping. An increase in collector doping offsets the Kirk effect as indicated by Equation (3.4), but again, a fundamental limit is reached when the doping becomes so high that reverse bias leakage between the base and collector, due to either avalanche multiplication or tunneling, dominates the device behavior. Due to the reasons discussed in Section 3.2, the additions of Ge and C into the base of the HBT serve to delay the point at which these fundamental limits are reached and thus allow further vertical scaling of the device than would be possible for a homo-junction device. It should be noted that scaling down of the base width or scaling up of the collector doping both serve to reduce f_{MAX} by increasing R_B in the first case and by increasing C_{BC} in the latter (see Equation (3.3)). Thus, most techniques employed to enhance f_T trade-off with a reduction in f_{MAX} and other techniques are needed to simultaneously improve f_{MAX}. Higher f_{MAX} values are enabled by lateral scaling of the devices. Smaller device dimensions serve to reduce the R_B and C_{BC} terms in Equation (3.3). In fact, most of the research involved in developing a new generation of SiGe HBT devices involves creating new ways to reduce these two parasitic parameters. At the heart of the scaling of SiGe HBT devices is the emitter width, which in turn limits most of the other dimensions in the device as a whole. While the most advanced SiGe HBT devices constructed to date have emitter widths less than 100 nm [9], scaling of the emitter width is roughly 10 years behind the scaling of CMOS gate lengths. Figure 3.5 shows projection data from the ITRS roadmaps for CMOS and bipolar technologies [10], showing projected f_{MAX} vs. the minimum feature width in the given technology node. It shows that, on average, one can achieve the same f_{MAX} in a bipolar device with a minimum feature width roughly three times larger than CMOS.

Several device architectures have been developed in the past decade in order to allow scaling of SiGe HBT devices down to nanoscale dimensions. Figure 3.6 shows a generic example of the device architecture used to construct most modern SiGe HBTs. The first large-scale manufactured SiGe HBT devices were built with a "quasi-self-aligned" architecture [11] where the extrinsic base is self-aligned by ion implantation to the edges of the emitter poly but not the emitter itself (see Figure 3.6(a)). The next generation of devices split off into several

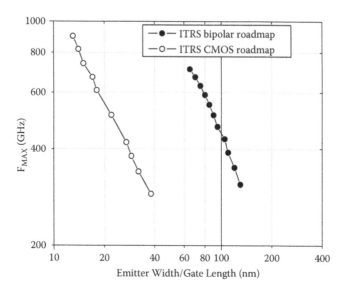

FIGURE 3.5 f_{MAX} vs. minimum feature size for bipolar vs. CMOS technologies. Data are from the ITRS roadmaps for CMOS and bipolar technologies. (From *2010 International Technology Roadmap for Semiconductors*, International SEMATECH, Austin, TX.)

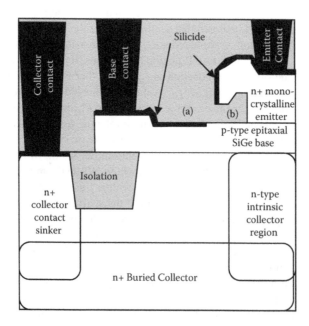

FIGURE 3.6 Example cross section of a modern SiGe HBT. (a) denotes the silicided extrinsic base region and the implanted extrinsic base region in quasi-self-aligned devices, and (b) denotes the "link" or "spacer" region, which is doped (and thus implies "fully self-aligned") by the various techniques discussed in the text.

different architectures that are "fully self-aligned," meaning that the extrinsic and intrinsic base alignment does not depend on mask alignment. One type of device uses a deposited polycrystalline extrinsic base followed by "selective epitaxy" of the intrinsic SiGe base [12]. A second method uses a sacrificial emitter post and spacer similar to the construction of a CMOS device [13]. Finally, various methods of growing a "raised extrinsic base" after the epitaxial growth of the intrinsic SiGe base have been developed [14, 15]. All of these modern techniques essentially serve to dope the region denoted in Figure 3.6(b) at a higher p-type level than that of the SiGe epitaxy. This extra doping in the extrinsic base region serves to lower the total R_B of the device and thus improve f_{MAX}. While the techniques used to enhance f_T discussed above tend to reduce f_{MAX}, the scaling and architecture enhancements discussed here serve to improve f_{MAX} without any significant penalty to f_T. Thus, these innovations have allowed continuous scaling of SiGe HBT devices akin to what is done in CMOS.

Figure 3.7 shows a compilation of f_T and f_{MAX} data from over 100 SiGe HBT publications overlaid with the 2010 ITRS roadmap for bipolar devices [10]. The scatter plot shows the basic correlation of the progression of f_T and f_{MAX} despite the trade-offs mentioned above. The roadmap data predict that the same lithography advancements responsible for the CMOS roadmap will enable improved SiGe performance for the foreseeable future.

To realize useful RF and analog circuits, however, more than just high-speed SiGe devices are necessary. In the next section, we will discuss modules integrated with SiGe transistors that help create a more complete modern platform for RF and analog IC design.

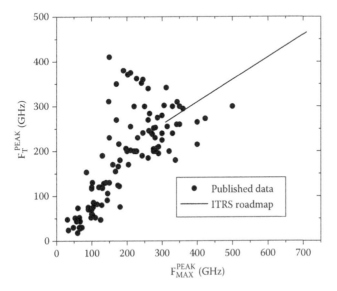

FIGURE 3.7 f_T vs. f_{MAX} scatter plot for published SiGe HBT data. The line is projection data from the ITRS bipolar roadmap. (From C. T. Kirk, *IRE Trans. Elec. Dev.*, 9(2), 164, 1962.)

3.6 MODERN SiGe BICMOS RF PLATFORM COMPONENTS

Technology features integrated with SiGe transistors that make them useful for product design include active elements such as high-density CMOS, high-voltage CMOS, high-performance PNP-type bipolar transistors, as well as passive elements such as high-density metal-insulator-metal (MIM) capacitors and high-quality inductors.

Today, most SiGe development is done in the context of a BiCMOS process in a CMOS node that typically trails the most advanced digital node by several generations. The critical hurdle to integrating advanced CMOS and SiGe devices is to marry their respective thermal budgets without degrading either device. The addition of carbon to SiGe layers as discussed in Section 3.2 has been used as a partial solution to this problem, as it helps reduce boron diffusion, allowing for a higher thermal budget after SiGe deposition. This, along with careful optimization of the integration scheme, has resulted in demonstrations of SiGe integration down to the 90 nm node [16].

Power management circuitry can be enabled with higher-voltage CMOS devices (typically requiring tolerance of 5 to 8 V). In smaller geometries that support only lower core voltage levels, these are enabled by introducing drain extensions to the CMOS devices that can enable higher drain bias than supported in the native transistor. An example of such devices is shown in Figure 3.8, and these

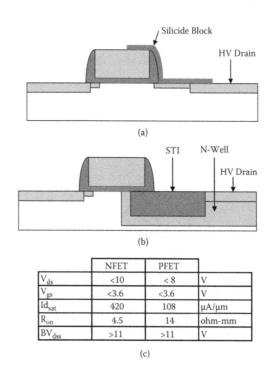

	NFET	PFET	
V_{ds}	<10	< 8	V
V_{gs}	<3.6	<3.6	V
Id_{sat}	420	108	μA/μm
R_{on}	4.5	14	ohm-mm
BV_{dss}	>11	>11	V

(c)

FIGURE 3.8 Sketch of two types of commonly used extended drain devices: (a) silicide block extension and (b) shallow trench isolation (STI) extension. (c) A table showing characteristics of high-voltage devices available in a 0.18 μm SiGe BiCMOS technology using approach (b). Idsat is quoted for 3.3 V Vgs and 5 V Vds.

are becoming common modules in SiGe technology offerings, often not costing additional masking layers to create.

A high-speed vertical PNP (VPNP) can form a complementary pair with the SiGe NPN bipolar transistor and is important for certain high-speed analog applications such as fast data converters, push-pull amplifiers, and output stages for hard disk drive preamps. A VPNP can be made very fast by the use of a separate SiGe deposition step, and f_T values as high as 100 GHz have been reported [17]. But the cost associated with such a VPNP is prohibitive for most applications today. A more popular approach reuses many of the steps needed to create the SiGe HBT and CMOS devices while adding specialized implants to optimize the performance of the VPNP. In this scheme devices with f_T values of up to 30 GHz can be achieved, as shown in Figure 3.9.

In addition to active components, high-quality passive components are necessary to enable advanced RF circuits. The most critical passive elements for RF design are capacitors and inductors, as these can consume significant die area and, at times, limit performance of RF and analog circuits. Metal-insulator-metal (MIM) capacitors are available in most commercial SiGe BiCMOS and RF CMOS processes, as they achieve excellent linearity and matching. The density of MIM capacitors has been steadily increasing over time, helping to shrink RF and analog die. Figure 3.10 shows a timeline of capacitance density for TowerJazz integrated MIM capacitors. An initial improvement in density from <1 fF/µm² to 2 fF/µm² was enabled by a move from oxide to nitride dielectrics [18]. Then, a move from 2 fF/µm² to 4 fF/µm² was enabled by the stacking of a 2 fF/µm² capacitor on two consecutive metal layers. Finally, a further optimization of the nitride dielectric resulted in a density of 5.6 fF/µm². Today, high-K dielectrics and various types of MIM trench capacitors

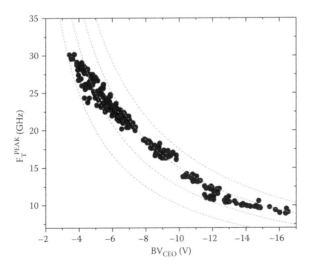

FIGURE 3.9 Performance (f_T) vs. breakdown (BV_{CEO}) trade-off of vertical silicon PNP devices integrated with SiGe NPNs to form a complementary pair. The dashed lines are contours of constant $f_T \times BV_{CEO}$. Data are from internal TowerJazz development wafers.

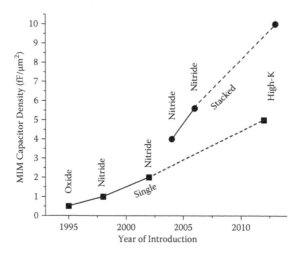

FIGURE 3.10 MIM capacitor density plotted as a function of year of first production (actual or planned) showing progression in dielectric technology (from oxide to nitride to high K) and in integration (single to stacked capacitors).

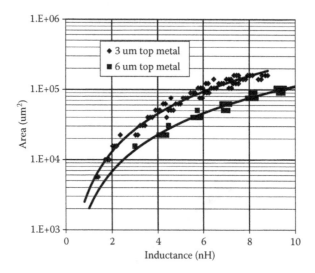

FIGURE 3.11 Inductor area as a function of inductance for a four-turn inductor with peak Q of 10 built in 3 and 6 μm Al metal layers, respectively.

are being investigated to enable even higher densities, and it is conceivable that in the next few years densities of 10 to 20 fF/μm² will be introduced.

Integrated inductor performance, measured as the quality factor (Q), is improved by the reduction of metal resistance made possible by thicker metal layers. Inductor Q can be traded off for reduced footprint such that a thicker metal layer can also help reduce chip area. This concept is demonstrated in Figure 3.11, where the area required to realize an inductor with Q of 10 is compared between use of a 6 μm and a 3 μm top metal in a four-layer Al metal process. A 6 μm metal inductor consumes half the die area of a 3 μm metal inductor while achieving the same Q in this example.

Die scaling enabled by the advanced passive elements described in this section can often more than pay for the additional processing cost. An optimized process can, in many cases, not only provide better performance than a digital CMOS process, but also lower the die cost. Similarly, the integration of advanced active modules described in this section can help integrate more analog functionality on fewer die, reducing overall system level costs.

3.7 CONCLUSIONS

In this chapter, we have reviewed SiGe BiCMOS technology and discussed how it has become important for many RF applications by providing a performance advantage over stand-alone CMOS while sharing its manufacturing infrastructure to provide integration and cost advantages over III-V technology. In addition to higher speed, we have seen that an intrinsic advantage of SiGe over CMOS is its ability to maintain higher breakdown voltages, and therefore support applications that require a higher dynamic range. This gap will widen with more advanced generations of both CMOS and SiGe, as each new generation of CMOS results in lower breakdown voltages, while each new generation of SiGe results in a better trade-off between speed and breakdown. In addition, we have seen that performance of SiGe devices can be improved with advanced lithography, much in the same way as with CMOS devices, such that a raw performance gap will continue to exist between SiGe and CMOS as more advanced nanometer nodes are created in the future. This will continue to enable a market for SiGe at the bleeding edge of performance, which today is translating into interest for SiGe in several millimeter-wave applications and very high-speed networks.

The biggest threat to SiGe advancements is the failure to identify high-speed, high-volume applications that take advantage of these benefits in the future, but much like Moore's law for CMOS has held true for decades and applications have taken full advantage, the imagination of the industry has never let us down before and is not likely to do so in this case.

ACKNOWLEDGMENTS

The authors acknowledge the help of current and former coworkers at TowerJazz, including Volker Blaschke, Dieter Dornisch, David Howard, Chun Hu, Paul Hurwitz, Amol Kalburge, Arjun Karroy, Paul Kempf, Lynn Lao, Zachary Lee, Pingxi Ma, Greg U' Ren, Jie Zheng, and Bob Zwingman, as well as contributions from many of TowerJazz's customers and partners. A final special acknowledgment is given to Jay John of Freescale Semiconductor for compiling the large dataset of published f_T and f_{MAX} data used to generate Figure 3.7.

REFERENCES

1. H. Kroemer, Heterostructure bipolar transistors and integrated circuits, *Proc. IEEE*, 70, 13–25, 1982.
2. D. L. Harame, J. H. Comfort, J. D. Cressler, E. F. Crabbe, J. Y.-C. Sun, B. S. Meyerson, and T. Tice, Si/SiGe epitaxial-base transistors. I. Materials, physics, and circuits, *IEEE Trans. Elec. Dev.*, 455–482, 1995.
3. L. D. Lanzerotti, J. C. Sturm, E. Stach, R. Hull, T. Buyuklimanli, and C. Magee, Suppression of boron outdiffusion in SiGe HBTs by carbon incorporation, *IEDM Tech. Dig.*, 249–252, 1996.
4. T. Manku and A. Nathan, Electron drift mobility model for devices based on unstrained and coherently strained $Si_{1-x}Ge_x$ grown on <001> silicon substrate, *IEEE Trans. Elec. Dev.*, 39(9), 2082–2089, 1992.
5. G. L. Patton, D. L. Harame, J. M. C. Stork, B. S. Meyerson, G. J. Scilla, and E. Ganin, Sige-base, poly-emitter heterojunction bipolar transistors, in *Symposium on VLSI Technology*, 1989, pp. 35–36.
6. IEEE 802.15.3c, IEEE 802.11ad.
7. C. T. Kirk, A theory of transistor cutoff frequency (fT) falloff at high current densities, *IRE Trans. Elec. Dev.*, 9(2), 164, 1962.
8. S. P. Voinigescu, M. C. Maliepaard, J. L. Showell, B. E. Babcock, D. Marchesan, M. Schroter, P. Schvan, and D. Harame, A scalable high-frequency noise model for bipolar transistors with application to optimal transistor sizing for low-noise amplifier design, *J. Solid-State Circuits*, 32(9), 1430–1439, 1997.
9. T. Lacave, P. Chevalier, Y. Campidelli, M. Buczko, L. Depoyan, L. Berthier, G. Avenier, C. Gaquiere, and A. Chantre, Vertical profile optimization for +400GHz fMAX Si/ SiGe:C HBTs, in *2010 BCTM Proceedings*, 2010, pp. 49–52.
10. *2010 international technology roadmap for semiconductors*, International SEMATECH, Austin, TX.
11. H. Miyakawa, M. Norishima, Y. Niitsu, H. Momose, and K. Maeguchi, A 3.3V, 0.5µm BiCMOS technology for BiNMOS and ECL gates, in *1991 CICC Proceedings*, 1991, pp. 18.3/1–18.3/4.
12. F. Sato, T. Hashimoto, T. Tashiro, T. Tatsumi, M. Hiroi, and T. Niino, A novel selective SiGe epitaxial growth technology for self-aligned HBTs, in *Symposium on VLSI Technology*, 1992, pp. 62–63.
13. M. Racanelli, K. Schuegraf, A. Kalburge, A. Kar-Roy, B. Shen, C. Hu, D. Chapek, D. Howard, D. Quon, F. Wang, G. U'Ren, L. Lao, H. Tu, J. Zheng, J. Zhang, K. Bell, K. Yin, P. Joshi, S. Akhtar, S. Vo, T. Lee, W. Shi, and P. Kempf, Ultra high speed SiGe NPN for advanced BiCMOS technology, in *2001 IEDM Proceedings*, 2001, pp. 15.3.1–15.3.4.
14. B. Jagannathan, M. Khater, F. Pagette, J.-S. Rieh, D. Angell, H. Chen, J. Florkey, F. Golan, D. R. Greenberg, R. Groves, S. J. Jeng, J. Johnson, E. Mengistu, K. T. Schonenberg, C. M. Schnabel, P. Smith, A. Stricker, D. Ahlgren, G. Freeman, K. Stein, and S. Subbana, Self-aligned SiGe NPN transistors with 285 GHz fMAX and 207 GHz fT in a manufacturable technology, *IEEE Elec. Dev. Lett.*, 23(5), 258–260, 2002.
15. H. Rucker, B. Heinemann, R. Barth, D. Bolze, J. Drews, U. Haak, W. Hoppner, D. Knoll, S. Marschmeyer, H. H. Richter, P. Schley, D. Schmidt, R. Scholz, B. Tillack, W. Winkler, H.-E. Wulf, and Y. Yamamoto, SiGe:C BiCMOS technology with 3.6ps gate delay, in *2003 BCTM Proceedings*, 2003, pp. 121–124.
16. K. Kuhn, M. Agostinelli, S. Ahmed, S. Chanbers, S. Cea, S. Christensen, P. Fischer, J. Gong, C. Kardas, T. Letson, L. Henning, A. Murthy, H. Muthali, B. Obradovic, P. Packan, S. W. Pae, I. Post, S. Putna, K. Raol, A. Roskowski, R. Soman, T. Thomas,

P. Vandervoorn, M. Weiss, and I. Young, A 90nm communication technology featuring SiGe HBT transistors, RF CMOS, precision R-L-C RF elements and 1 μm^2 6-T SRAM cell, *IEDM Tech. Dig.*, 73–76, 2002.

17. B. Heinemann, R. Barth, D. Bolze, J. Drews, P. Formanek, O. Fursenko, M. Glante, K. Lowatzki, A. Gregor, U. Haak, W. Hoppner, D. Knoll, R. Kurps, S. Marschmeyer, S. Orlowski, H. Rucker, P. Schley, D. Schmidt, R. Scholz, W. Winkler, and Y. Yamamoto, A complementary BiCMOS technology with high speed npn and pnp SiGe:C HBTs, *IEDM Tech. Dig.*, 117–120, 2003.

18. A. Kar-Roy, C. Hu, M. Racanelli, C. A. Compton, P. Kempf, G. Jolly, P. N. Sherman, J. Zheng, Z. Zhang, and A. Yin, High density metal insulator metal capacitors using PECVD nitride for mixed signal and RF circuits, in *Interconnect Technology IEEE International Conference*, 1999, pp. 245–247.

4 SiGe HBT Technology and Circuits for THz Applications

Jae-Sung Rieh

CONTENTS

4.1   INTRODUCTION

The SiGe heterojunction bipolar transistor (HBT) is basically a Si-based bipolar junction (BJT) transistor with a small amount of Ge added to the base region. The SiGe base region turned out to be a significant performance improver, making SiGe HBTs now accepted as a standard bipolar transistor for high-speed applications. When first demonstrated in 1987 by researchers at IBM based on molecular beam epitaxy (MBE)-grown SiGe/Si epitaxial layers [1], SiGe HBTs were in fact intended for high-speed digital applications. However, partly due to the huge power dissipation when highly integrated and also due to the concurrent emergence of complementary metal-oxide-semiconductor (CMOS) circuits with low-power operation, SiGe HBTs had to redefine their application path. It was fortunate for the devices to find a new fertile application area, namely, the radio frequency (RF) circuits and systems. The first report on the RF characteristics of SiGe HBT were brought out in 1989 by a group of researchers at Stanford and Hewlett-Packard, who exhibited an f_T of 28 GHz [2]. Since then, a series of significant improvements in the RF performance followed, notably the one achieving the first f_T exceeding 100 GHz in 1993 [3]. Another great milestone from a practical point of view was the release of the world's first commercial SiGe HBT technology (in the form of a BiCMOS technology) by IBM in 1996, which exhibited f_T and f_{max} of 47 GHz and 65 GHz, respectively [4]. Continuing scaling efforts, combined with structural innovation such as raised extrinsic base, led to SiGe HBTs operating beyond 200 GHz based on a 0.13 μm lithography node [5], which was soon followed by a device with an f_T of 375 GHz [6]. With the abundant circuit and system applications of SiGe HBTs that benefited from the technology development, it became obvious for technology developers that f_{max} is more desired than f_T from the circuit designers' side. Consequently, the technology innovation made a small change in its roadmap in more favor of high f_{max} of the device. With such efforts, SiGe HBTs with f_{max} higher than 400 GHz were reported by STM [7], and soon after, a 500 GHz f_{max} SiGe HBT was released by IHP [8], owing to the organized efforts to achieve a half-terahertz operation as supported by the Dot Five project [9]. The recent technological advancements in SiGe HBTs significantly contributed to the emergence of the Si-based THz era.

It is true that the performance improvement of Si metal-oxide-semiconductor field-effect transistors (MOSFETs) during the same period was remarkable too, owing to the aggressive scaling driven by the needs from mainstream digital application, leading to the successful adoption of Si CMOS technology for RF applications. However, SiGe HBTs still remain a favored option for RF applications from various aspects. The transconductance and output impedance, essential for sufficient voltage gain required for various circuit blocks, favor bipolar transistors due to their exponential current-voltage relation and graded bandgap in the base. The vertical nature of the bipolar structure results in excellent low-frequency noise as the current path stays away from the trap-rich interfaces. Device matching also benefits from its vertical structure as the vertical dimension is typically better controlled than the lateral dimension in modern semiconductor fabrication processes. Probably the most important aspect is that the performance of a bipolar transistor is not much driven by lateral scaling as in the case for Si MOSFETs, but dictated by the vertical scaling. Accordingly, a bipolar transistor can exhibit performance similar to that of Si MOSFETs with a few lithography nodes ahead. Considering the high cost required for every lithography node advance, in both masks and lithography tools, such a relaxed demand on the lateral scaling saves lots of cost for SiGe HBTs.

In this chapter, an overview of device and circuit aspects of SiGe HBT technology is presented. In Section 4.2, the device aspects of SiGe HBTs will be described, followed by a review of recent circuit implementation based on SiGe HBT technologies toward THz application in Section 4.3.

4.2 SiGe HBTs: DEVICE ASPECTS

4.2.1 DC Characteristics

As briefly mentioned earlier, the distinguishing feature of SiGe HBTs from the conventional Si bipolar junction transistors is the incorporation of the SiGe layer in the base region. A cross section of a typical SiGe HBT is depicted in Figure 4.1, which

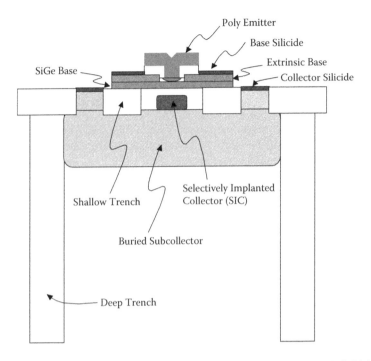

FIGURE 4.1 Schematic cross section of a typical modern SiGe HBT. (From J.-S. Rieh et al., *IEEE Transactions on Microwave Theory and Techniques*, 52, 2390–2408, 2004.)

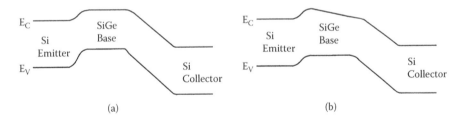

FIGURE 4.2 Band diagram of a SiGe HBT biased with forward active mode: (a) uniform Ge composition and (b) linearly retrograded Ge composition.

was developed by IBM. The inclusion of Ge in the base region leads to heterojunctions established between the emitter and base as well as the base and collector, as the bandgap of the SiGe layer is smaller than that of Si. The band diagrams along the vertical cut line in the intrinsic region of the device are shown in Figure 4.2. A direct consequence arising from the heterojunction is the significantly increased current gain of the device. It is noted that for the forward active operation of the device, the critical junction that affects the current gain is the E-B junction rather than the B-C junction. The increase in the current gain is in fact ascribed to the increased emitter current with the reduced base bandgap. It is interesting to note that the base current is not affected by the base bandgap reduction. The current gain enhancement with

SiGe HBTs over the conventional Si BJTs exponentially increases with the bandgap difference between the emitter and base, ΔE_{EB}, as shown below for the case of a uniform Ge composition across the base region:

$$\frac{\beta_{SiGe,unif}}{\beta_{Si}} \approx e^{\Delta E_{EB}/kT} \tag{4.1}$$

where $\beta_{SiGe,unif}$ and β_{Si} represent the current gain of SiGe HBT and Si BJT, respectively. It is assumed that the density of states and electron mobility in the base region are roughly the same for the two devices, while a more exact expression can be found elsewhere [10]. With the typical range of Ge composition of 10–20%, the current gain enhancement can range from ~35 to as high as ~900 at room temperature. The number can be even greater at lower temperatures. A typical modern SiGe HBT employs a retrograded Ge composition (higher Ge composition at the collector side edge of base) rather than uniform composition to exploit the quasi-electric field in the base region. In this case, the current gain enhancement factor is slightly modified as follows if a linearly graded profile is assumed:

$$\frac{\beta_{SiGe,grade}}{\beta_{Si}} \approx e^{\Delta E_{EB0}/kT} \frac{\Delta E_{grade}}{kT} \tag{4.2}$$

where ΔE_{EB0} is the bandgap difference between the emitter region and the emitter-side edge of the neutral base region (i.e., undepleted base region), while ΔE_{grade} represents the bandgap change across the neutral base region. When compared to the uniform Ge composition case, there is an additional linear factor due to ΔE_{grade} for the current gain enhancement. If a grading of 10% is assumed, the additional factor would be 3.5, further enhancing the current gain. It is noted, though, that there is a chance that ΔE_{EB0} is smaller than ΔE_{EB}, especially when the total amount of Ge included in the base is approximately the same for the two cases being compared.

Based on the analysis, the current gain of a SiGe HBT can easily rise beyond several thousands or even up to several tens of thousands if a typical doping profile of Si BJTs is maintained. However, the excessively large current gain is not necessarily favored, as it can lead to degradation in the breakdown voltages without additional benefits for circuit applications. A rule of thumb is that the current gain of about 100 is sufficient for most circuit applications. Any extra current gain larger than this will not be much desired. The real advantage in the device design of SiGe HBTs, however, lies in the fact that the extra current gain can be traded for the benefit of RF characteristics of the device. A typically practiced trade-off is to increase the base doping concentration in favor of enhanced operation speed. The increased base doping will lower the current gain, but will provide benefits such as base resistance reduction and further reduced base layer width without emitter-collector punch-through. These aspects will significantly improve the RF performance of the device, as will be detailed shortly.

Another favored feature of SiGe HBTs compared to Si BJTs in their DC characteristics is the increased Early voltage V_A. The Early voltage, defined as the inverse of the fractional change of collector current over V_{CB} variation, is a measure of the sensitivity of the collector current to the modulation of collector-base bias V_{CB}. Hence, V_A is an indicator of the output conductance of the device. The mechanism of the Early effect is as follows. The collector current is affected by V_{CB} variation because the neutral base thickness tends to decrease with increasing V_{CB} as it causes the B-C depletion width to increase. Reduced base thickness leads to increase in emitter current and thus collector current. There are a couple of reasons behind the larger V_A observed for SiGe HBTs. The large base doping concentration of SiGe HBTs makes the neutral base thickness rather insensitive to V_{CB} variation. Additionally, for the SiGe HBTs with retrograded base, the collector current is less sensitive to the change in the location of the collector-side edge of the neutral base. Large V_A of SiGe HBTs is translated into large output impedance, leading to higher voltage gain of amplifiers for a given load.

Generally, semiconductors with smaller bandgap tend to show lower breakdown voltages. As the bandgap of SiGe is smaller than that of Si, it is rational to presume that SiGe HBTs would exhibit smaller breakdown voltages than Si BJTs. In fact, such conjecture is not true for B-C junction and only partially true for E-B junction of SiGe HBTs due to the following reasons. For B-C junction, the depletion region is formed mostly in the Si collector side as the collector doping is typically much lower than the base doping. Hence, the breakdown will mostly occur in the Si collector region of the B-C junction, exhibiting a similar breakdown behavior as Si BJTs. For the E-B junction, the details will depend on the relative level of emitter and base doping concentration. If the base is more heavily doped than the emitter, the depletion will occur mostly in the Si emitter side and the behavior will be similar to the Si BJT case, while the breakdown will be affected by the narrow bandgap of SiGe if emitter doping is higher. For most SiGe circuit applications, the device will be operating in the forward active mode, in which the B-C junction is reverse biased while the E-B junction is forward biased. Hence, it is the B-C junction breakdown that is more relevant from a practical perspective and the narrow bandgap of the SiGe base is not a liability for breakdown performance of SiGe HBTs.

Nevertheless, it is still true that the modern high-speed SiGe HBTs tend to show smaller breakdown voltages than traditional Si BJTs. In fact, such tendency is not because of the SiGe base, but because of the more aggressive doping profile of SiGe HBTs intended for high-speed operation. As will be discussed later, higher collector doping concentration is favored for enhanced device operation speed, and thus eagerly adopted for SiGe HBTs. Such design, however, necessarily demands sacrifice of the breakdown voltage. Most advanced SiGe HBTs of today typically show BV_{CEO} of around 1.5 V. However, it is important to note that the value is not necessarily the upper limit of the voltage allowed across the emitter and collector in actual circuit applications. This becomes more obvious if we examine the various definitions of E-C breakdown voltage: BV_{CEO}, BV_{CES}, and BV_{CER}. BV_{CEO} is defined as the E-C breakdown voltage with the base open. This is the worst configuration for E-C breakdown voltage since the generated electrons by junction breakdown cannot exit through the base, and end up accelerating the breakdown mechanism. The opposite extreme is BV_{CES}, which is the E-C breakdown voltage with base shorted to a grounded emitter. In this case, the generated electrons easily find exit to ground and the junction breakdown mechanism is basically the same as the B-C junction breakdown. A general case is represented by BV_{CER}, which is defined as the E-C breakdown voltage with base connected to grounded emitter through finite resistance, and should have a value in between BV_{CES} and BV_{CEO}. This configuration is the case for most practical circuit situations, and thus BV_{CER} best represents the maximum allowed voltage across the emitter and collector, the actual value being dependent on the effective resistance between the base and collector. A typical BV_{CES} for most advanced SiGe HBTs of today is around 5 V. Hence, in most cases, voltages substantially higher than BV_{CEO} can be allowed across the emitter and collector.

4.2.2   RF CHARACTERISTICS

The chief advantage of SiGe HBTs over Si BJTs is the superior high-frequency characteristics. There are two widely used measures for the high-frequency performance of transistors: the cutoff frequency f_T and the maximum oscillation frequency f_{max}, which are defined as the frequency where the current gain and the unilateral power gain become unity, respectively. The relation between device parameters and each of these performance measures is discussed below.

The cutoff frequency f_T has an inverse relation with the emitter-to-collector delay time, τ_{EC}. Thus, it is useful to express τ_{EC} as a function of transistor parameters as shown below:

$$\tau_{EC} = \frac{1}{2\pi f_T} = \tau_E + \tau_C + \tau_B + \tau_{CSCL}$$

$$= \frac{kT}{qI_C}C_{EB} + \left(\frac{kT}{qI_C} + R_C + R_E\right)C_{CB} + \frac{W_B^2}{\gamma' D_n} + \frac{W_{CSCL}}{2v_{SAT}}$$

(4.3)

where k is the Boltzmann constant, C_{EB} and C_{CB} are E-B and B-C capacitances, R_C and R_E are collector and emitter resistances, W_B is the neutral base width, W_{CSCL} is the B-C space-charge region (SCR) width, D_n is the electron diffusion constant, v_{SAT} is the saturation velocity, and γ' is the quasi-electric field factor. The major effect of employing Ge in the base region on the cutoff frequency is the reduction of the base transit time, or τ_B. There are two main factors that contribute to the reduction of τ_B by employing Ge in the base region for SiGe HBTs. The first factor is the availability of aggressively scaled W_B by trading the extra current gain with increased base doping as mentioned earlier. With low base doping, W_B cannot be reduced too much, as it will trigger the emitter-collector punch-through. The achieved narrow base will directly result in the reduction of τ_B. The second factor is the quasi-electric field established across the base region when graded Ge composition is employed. With a retrograded Ge composition, energy bandgap in the base region will gradually decrease toward the collector as shown in Figure 4.2(b), accelerating the injected electrons with a quasi-electric field established in the base region. The quasi-electric field can reach as high as ~10 kV/cm, assuming 10% Ge grading across a 100 nm neutral base width, reducing the base transit time by a factor of ~3.

Collector scaling is another popular strategy often practiced in industry for performance enhancement. The primary effect of the collector scaling is reduced W_{CSCL}, which is usually achieved with increased collector doping, resulting in a reduction in the transit time across the B-C space-charge region, τ_{CSCL}. The reduction of W_{CSCL}, however, necessarily involves an increase in C_{CB}, leading to increased collector delay, τ_C. Therefore, a precise balance between these two conflicting aspects is required in collector scaling, a major engineering challenge for SiGe HBT optimization. One common approach is to adopt a selectively implanted collector (SIC), for which only the central intrinsic region of the collector is doped, while the extrinsic region remains undoped. This will effectively suppress the extrinsic component of C_{CB}, while reducing the transit time across the B-C space-charge region. In addition to τ_{CSCL} reduction, another important effect of increased collector doping is the increased Kirk current I_K, defined as the collector current at which the Kirk effect takes place. f_T of bipolar transistors tends to increase with increasing collector current for low current regime owing to the reduced charging time, as shown in Figure 4.3, which depicts the RF characteristics of a typical modern SiGe HBT. However, when the collector reaches around I_K, f_T begins to show an abrupt drop due to the Kirk effect. Hence, large I_K is desired to obtain high peak f_T, which is achieved with increased collector doping concentration. Note that I_K is roughly around the current level that shows the peak f_T, which is denoted as I_{peak} in Figure 4.3. Yet another effect of the collector scaling is an increase in the average velocity across the B-C space-charge region due to enhanced ballistic transport [11], leading to reduced τ_{CSCL}. However, the collector scaling inevitably results in the lowered breakdown voltages. Hence, there exists a trade-off between the operation speed and the breakdown voltage of the device. Such a trade-off is rather fundamental, and an upper limit for the product of the breakdown voltage and the cutoff frequency may exist, which has long been known as the Johnson limit [12]. Although the Johnson limit turned out to be pessimistic and has now been overcome with modern technology [13], it is still true that there exists such a trade-off, another design point that needs to be decided for SiGe HBT technology development.

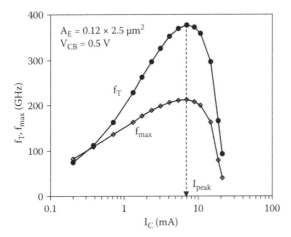

FIGURE 4.3   RF characteristics of a typical modern SiGe HBT in terms of f_T and f_{max}.

Both base and collector scalings considered so far are basically the vertical scaling. Lateral scaling, which is a critical element for MOSFET performance enhancement, also imposes effects on f_T of bipolar transistors. The increase of C_{CB} resulting from the vertical scaling can be partly compensated by the lateral scaling. C_{EB} can be reduced by the lateral scaling as well. On the other hand, lateral dimension reduction will lead to the degradation of R_E and R_C as a result of the laterally narrowed vertical current paths. The fringing current component due to an increased emitter perimeter-to-area ratio also tends to grow when the emitter stripe width is decreased. Hence, it is fair to say that the impact of the lateral scaling on f_T is relatively limited for bipolar transistors.

Another measure of device speed, f_{max}, is considered to be a more relevant parameter to assess the high-frequency performance of transistors in most circuit applications as mentioned earlier. A simplified expression for f_{max} of bipolar transistors is given as follows:

$$f_{\max} = \sqrt{\frac{f_T}{8\pi R_B C_{CB}}} \qquad (4.4)$$

As can be seen from Equation 4.4, f_{max} improves with f_T, but is also affected by R_B and C_{CB}. The effect of the vertical scaling on f_{max} of SiGe HBTs is not straightforward. f_T is clearly improved with the vertical scaling, which helps to improve f_{max} as well according to Equation 4.4. On the other hand, R_B and C_{CB} can be degraded by the vertical scaling. Base current is supplied laterally to the intrinsic device, and the vertical base scaling will inevitably pinch the lateral path for base current, leading to increased R_B. The vertical collector scaling will narrow down the B-C space-charge layer in the intrinsic portion of the device, leading to increased C_{CB}. As a result, the impact of the vertical scaling on f_{max} is complicated and may turn out either positive or negative depending on the structural details of a given device. In contrast, the lateral scaling, which has only minor effects on f_T, as discussed above, plays a major role in f_{max}. The lateral scaling shortens the resistive current path for base, resulting in R_B reduction. The B-C junction area decreases with lateral scaling, effectively reducing C_{CB}. The lowered R_B and C_{CB} with lateral scaling will contribute to boost f_{max}.

Up to this point, the effects of the vertical and lateral scalings on f_T and f_{max} of SiGe HBTs have been discussed. Another design point that significantly affects the device characteristics is the device layout. The effect of the layout configuration on RF performance is illustrated in Figure 4.4 [14]. The three layout variants to be compared are denoted as CBE, CBEB, and CBEBC, which represent the relative locations of the emitter, base, and collector contacts. Si-based devices have traditionally adopted the compact CBE configuration since the contact and spreading resistance are not performance-limiting factors because of the availability of silicide. As the Si devices enter the hundred-GHz operation regime, however, the effect of parasitic resistance becomes more significant, and any layout variation that brings performance enhancement needs to be considered. Measurements show that, compared to the CBE case, the CBEB configuration improves f_{max} due to the reduction of the R_B component along the silicided region, while imposing little effect on f_T. On the other hand, CBEBC leads to the increase in both f_T and f_{max}, which can be ascribed

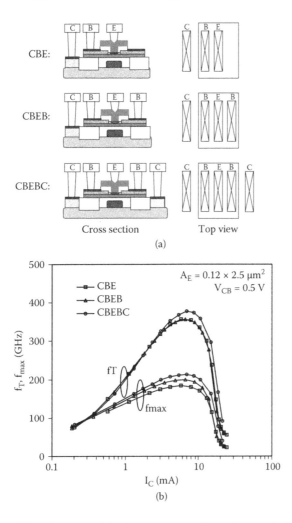

FIGURE 4.4   (a) The schematic of the three layout configurations typically employed for SiGe HBTs: CBE, CBEB, and CBEBC. (b) The effect of layout configuration on f_T and f_{max}. (From J.-S. Rieh et al., *IEEE Transactions on Microwave Theory and Techniques*, 52, 2390–2408, 2004.)

to *RC* delay reduction and the symmetric spread of injected electrons at the collector region that effectively delays the onset of the Kirk effect.

The dependence of peak f_T and f_{max} on emitter length L_E and width W_E is also presented in Figure 4.5 with CBE devices as an example [14]. The most evident trend for the L_E dependence is the increase of f_{max} with decreasing L_E, which can be ascribed to the decreasing R_B with the shorter base current path to the opposite side of the emitter finger along the silicided extrinsic base. Such an effect may be different for CBEB and CBEBC devices, though. f_T shows a weaker dependence on L_E than f_{max}, but it exhibits a moderate optimum point near $L_E = 2.5$ µm. This can be attributed to the competing scaling behavior of the resistance (R_E and R_C) and capacitance (C_{CB})

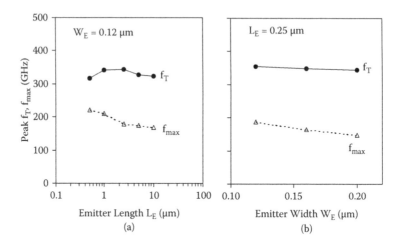

FIGURE 4.5 The dependence of peak f_T and f_{max} on emitter length L_E and width W_E with CBE devices. (From J.-S. Rieh et al., *IEEE Transactions on Microwave Theory and Techniques*, 52, 2390–2408, 2004.)

with L_E variation, an apparent balance occurring around this point. The reduction of f_T for smaller L_E is also caused by the increased peripheral component of injected electrons, which travels a longer path to the collector, thus resulting in a larger transit time than the area component of injected electrons. The impact of W_E modulation is also more pronounced on f_{max} than f_T, exhibiting a substantial increase of f_{max} with W_E reduction. This results from the reduced base spreading resistance underneath the emitter opening. The small dependence of peak f_T on W_E variation indicates that a balance is established between reduced C_{CB} and C_{EB} and increased R_C and R_E with the scaling, leaving the RC delay mostly unchanged.

4.2.3 NOISE CHARACTERISTICS

The noise properties of any active device can be represented by four noise parameters: minimum noise figure (F_{min}), noise resistance (R_n), and the real and imaginary components of the source impedance match for lowest F_{min} (Γ_{opt}). Among these parameters, F_{min} represents the lowest available noise level, and thus is widely used to assess the noise performance of a given device. For bipolar transistors, F_{min} can be expressed as a function of device parameters as follows:

$$F_{min} \simeq 1 + \sqrt{\frac{2qI_C}{kT} R_B \left(\frac{f^2}{f_T^2} + \frac{1}{\beta} \right)} \qquad (4.5)$$

For sufficiently high frequency, it is reduced to

$$F_{min} \simeq 1 + \frac{f}{f_T} \sqrt{\frac{2qI_C R_B}{kT}} \qquad (4.6)$$

Based on Equation 4.6, it is obvious that high f_T as well as small R_B helps to reduce the noise level of bipolar transistors for a given bias and frequency. Such a relation reveals that SiGe HBTs behave more favorably than Si BJTs in terms of noise performance as well as due to higher f_T and lower R_B typically available for SiGe HBT. It is also noted that the lateral scaling is helpful for noise performance, as in the case of f_{max}, since smaller W_E will yield smaller R_B as the resistive lateral path for base current beneath the emitter region is reduced. Typical high-frequency noise characteristics achieved with modern SiGe HBTs are shown in Figure 4.6 [14], in which F_{min} values from different generations of SiGe HBTs from IBM are plotted as a function of frequency. It shows a clear reduction in F_{min} between generations, mostly driven by the f_T improvement. Also shown for comparison is the typical range of F_{min} obtained from GaAs pseudomorphic high electron mobility transistors (PHEMTs), which are generally considered to show excellent high-frequency noise properties. As is clear from the figure, F_{min} from 200 GHz SiGe HBT is comparable to those from GaAs PHEMT, indicating the competent high-frequency noise characteristics of SiGe HBTs.

Low-frequency noise is another type of noise that affects the high-frequency characteristics of the device. Although there are several sources of low-frequency noise, the most dominant in modern HBTs is the $1/f$ noise, which stems from carrier number fluctuation by several trap-related mechanisms [15–18]. At medium to high bias, carriers are trapped and released by defects that are typically spread across emitter area A_E, concentrated near the thin interfacial oxide layer at the polysilicon-to-silicon emitter interface. A typical characteristic of $1/f$ noise of SiGe HBT is presented in Figure 4.7 for a device with $f_T = 120$ GHz, with a difference V_{BE} bias level. It shows a corner frequency, a typically adopted measure of the $1/f$ noise, of around 400 Hz, which is one to three orders of magnitude smaller than those of typical CMOS devices.

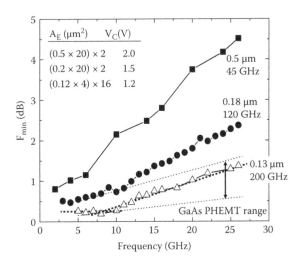

FIGURE 4.6 Typical high-frequency noise characteristics achieved with modern SiGe HBTs. (From J.-S. Rieh et al., *Proceedings of the IEEE*, 93, 1522–1538, 2005.)

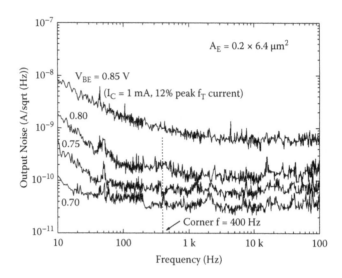

FIGURE 4.7 A typical characteristic of 1/*f* noise of SiGe HBT. (From J.-S. Rieh et al., *IEEE Transactions on Microwave Theory and Techniques*, 52, 2390–2408, 2004.)

4.2.4 FABRICATION PROCESS

Most of the major semiconductor companies that had owned established Si BJT fabrication process technology have developed SiGe HBT technology, modifying the existing BJT technology. The fabrication process of SiGe HBTs is largely similar to that of standard Si BJTs, the major difference being the formation of the base region that is formed by epitaxial growth of the SiGe alloy layer. The details of the fabrication process and completed device structure vary over different companies, but they share lots of basic schemes for the technology. A typical fabrication process of SiGe HBT is introduced here based on one of the recent technologies developed by IBM as an example [19], whose completed device structure is presented in Figure 4.1.

The process starts with the formation of a heavily arsenic-doped subcollector buried layer formed by ion implantation. On top of the implanted surface, a lightly doped epitaxial Si layer is grown that serves as the collector region. Two types of trenches, deep and shallow, are subsequently formed. Shallow trenches are the same as those used for CMOS device isolation, and serve to separate the active region and collector plug-in region of the device. Deep trenches are unique for bipolar transistors, which provide the primary isolation between adjacent devices. A highly doped collector pedestal region is next formed by a selective implantation into the collector layer, the doping of which is confined only within the central part of the active region. Such a structure, widely called selectively implanted collector (SIC), provides a small W_{CSCL} and high collector current density, while suppressing the increase in extrinsic C_{CB} by the implantation. Following is the formation of the boron-doped SiGe base layer, a key difference from the Si BJT processes. This step is critical for device performance, and thus requires a precise control of the Ge and boron doping profiles. It is a common practice to include a small amount of C in the base region, which is known to suppress the boron out-diffusion that otherwise would result in increased W_B. A self-aligned raised extrinsic base decouples extrinsic R_B and B-C overlap capacitance, providing a great flexibility in structure optimization for parasitic reduction. A heavily phosphorus-doped T-shaped polyemitter allows narrow emitter stripes close to 0.1 μm without causing a significant increase in emitter resistance. The final step for device formation is the silicidation of the extrinsic base and collector, which greatly reduces the contact resistance for the base and collector.

There may be some variations in the fabrication process steps depending on the final aim of the technology. For the devices intended for high-power applications, for example, the doping level of SIC can be reduced to increase the breakdown voltage, which will sacrifice the speed. When low cost is desired, the collector epitaxy step can be omitted, which will instead require high-energy implantation for subcollector formation. Deep trenches can also be eliminated or simplified for cost reduction as the formation of the high aspect ratio trench is costly and time-consuming. Both low-cost options will inevitably result in performance degradation, though, which is a classical trade-off between performance and cost. Some technologies employ a selective epitaxy process for the base layer formation in favor of the independent optimization of the extrinsic base, which will affect the base resistance and thus f_{max} as well as noise. It is finally noted that the HBT process can be integrated with the CMOS process, which is called the BiCMOS process. It is more favored for most applications that need integrated CMOS circuit blocks for various purposes, such as digital function. For BiCMOS fabrication processes, it is generally required that the process steps for HBT formation do not affect the performance of CMOS devices, often imposing an upper limit on the process temperature for HBT, which prevents the full optimization of the process flow for HBT.

The performance trend of SiGe HBTs is shown in Figure 4.8 in terms of f_T and f_{max} for selected technologies. As indicated by the plot, f_{max} has now exceeded 500 GHz [8], while f_T has also reached the 400 GHz level [20]. The device with

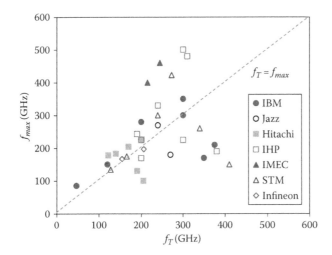

FIGURE 4.8 The trend of f_T and f_{max} of SiGe HBTs from selected foundries.

500 GHz f_{max} was developed by IHP as part of the Dot Five project, meeting the target frequency as suggested by the project title. The device is based on IHP's 0.25 μm BiCMOS technology environment with the emitter width of 0.12 μm. Such performance was made possible with various material and structural optimizations without drastic change from the conventional SiGe HBT processes, which includes the reduced thermal budget to suppress the boron out-diffusion, smaller emitter-base spacer width to reduce the extrinsic base resistance, and SIC implant before the base layer formation. One notable structural feature of this device is the removal of the shallow trench isolation between the active region and the collector plug-in, as depicted in Figure 4.9(a). The device showed BV_{CEO} and BV_{CBO} of 1.6 V and 5.2 V, respectively, and the emitter current density for the best speed performance was slightly smaller than 20 mA/μm². The associate f_T was around 300 GHz. f_{max} and f_T of the device are shown in Figure 4.9(b) as a function of collector current density, together with results obtained with older generations.

The achieved performance obtained so far with SiGe HBTs has certainly not reached the ultimate performance limit of the device yet. The limitation on f_T largely

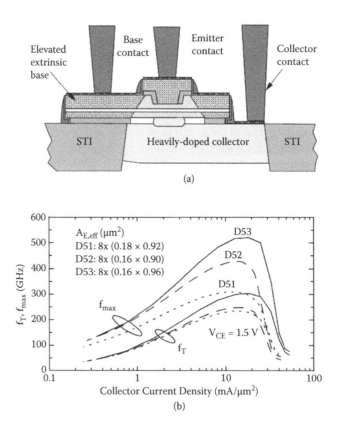

FIGURE 4.9 (a) The schematic cross section of the SiGe HBT with $f_{max} = 500$ GHz. (b) f_T and f_{max} of the SiGe HBTs recently developed in IHP. (From B. Heinemann et al., in *IEEE International Electron Devices Meeting*, 2010, pp. 30.5.1–30.5.4.)

comes from the constraints in the vertical scaling of the device, mostly those of doping and Ge composition. On the other hand, the limitation on f_{max} is additionally associated with the lateral scaling that reflects the effect of parasitics arising from the actual fabrication of the device. There have been numerous efforts to predict the ultimate performance limit of SiGe HBTs based on various simulations [21–25]. The most recent prediction with 1D simulation suggests that the ultimate value of f_T would fall upon around 1.5 THz, with BV_{CEO} of ~1 V [24]. When lateral effects are considered, including the parasitics based on 3D simulation, an f_{max} around 2 THz is expected to be feasible with associate f_T around 1 THz [25]. Although lots of barriers arising from the practical implementation issues need to be properly addressed, such prediction reveals that there still is much room for further performance enhancement, encouraging the efforts for THz application of SiGe HBTs.

4.3   SiGe HBTs: CIRCUIT APPLICATIONS

The achieved operation speed of a few hundred GHz with SiGe HBTs allows the development of circuits operating beyond 100 GHz, or the THz region. In fact, there are a growing number of reports on circuits based on SiGe HBTs that operate at this region, and they can be adopted for various THz applications, including broadband communication systems and imagers. Many of these applications need various circuit components, such as amplifiers, mixers, oscillators, frequency dividers, frequency multipliers, and so forth. In this section, I give an overview of recently reported circuits based on SiGe HBT, with a focus on these kinds of circuits together with integrated systems based on these unit circuit blocks.

4.3.1   AMPLIFIERS

The principal advantage of employing transistors over diodes for implementing THz systems is the availability of gain, which is realized with amplifiers. Amplifiers play a great role in various places in high-frequency systems. For receivers or detectors, a low-noise amplifier (LNA) is typically employed at the very first stage, which will significantly lower the overall noise level of the system. For transmitters or signal sources, a power amplifier (PA) is placed at the last stage for the final boost of output power transmitted, which is required to achieve a high signal-to-noise ratio at the receiver or detector side. Other places that benefit from amplifiers include the frequency multiplier chains, which are often employed in THz systems to elevate the frequency to the required level for transmitters or local oscillator blocks. As the signal power drops significantly after going through frequency multipliers due to their high conversion loss, amplifiers are typically inserted in the multiplier chain to compensate for the power loss.

Compared to other circuits such as mixers and oscillators, amplifiers demand more stringent requirements for the transistor operation frequency since a sufficient gain needs to be guaranteed at the required frequency. A simple analysis clearly shows such a challenge. A typically desired gain per stage in high-frequency amplifiers is around several dB. Considering the typical transistor RF characteristics that exhibit a gain roll-off of −6 dB/octave, which is basically a single-pole behavior, the required f_{max} for the device is about twice the operation frequency of the amplifier to meet the several dB gain per stage. In addition, there are usually additional loss factors for amplifiers, such as the loss of the passive components used for matching and possible mismatches arising from the inaccurate device model or electromagnet (EM) simulation, as well as process variation. Considering all of these factors, the desired f_{max} would be even higher.

There have been numerous reports on amplifiers based on SiGe HBTs operating beyond 100 GHz, which is shown in Figure 4.10(a) in terms of their gain vs. frequency characteristics. Also shown in Figure 4.10(b) is the gain per stage of the same set of amplifiers, which ranges between 3 and 11 dB, with typical values falling on around 5 dB. A measured noise figure is rarely reported in this range of frequency due to the limitation in the noise test setup. Simulations show that the noise figure (NF) of the amplifiers operating in this frequency band is expected to range between 8 and 11 dB, as shown in Figure 4.11, which is vulnerable to the accuracy of the noise model used for the process design kit (PDK). One measured NF reported for a 130 GHz amplifier is 6.8 dB, largely in line with the simulated values [26].

The highest operation frequency of SiGe amplifiers reported so far is around 250 GHz, which is based on IHP 0.25 μm SiGe BiCMOS technology with f_T/f_{max} of 300/500 GHz [27]. It employs a four-stage common base (CB) configuration, resulting in 12 dB overall gain, or 3 dB per stage. It is noted that the CB configuration typically shows higher gain than the common emitter (CE) configuration at this high frequency band owing to the inherently smaller C_{EB} than C_{BC} in bipolar transistors, which results in a greatly suppressed Miller effect. NF for this circuit is simulated to be around 11 dB. With a 2 V supply voltage, the amplifier consumes 28 mW DC power. The performance of this amplifier can be mostly attributed to the high f_{max} of the device employed. It is noted that the ideal gain available per stage would be around 6 dB at this frequency with the given device f_{max}, which is compared to the actually achieved gain per stage of 3 dB, exhibiting the challenge in the amplifier design at the THz region.

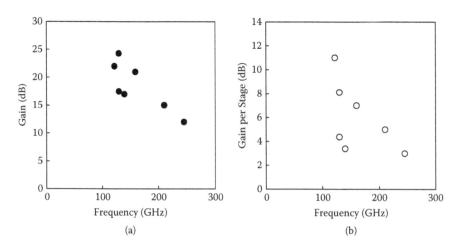

FIGURE 4.10 (a) Gain vs. frequency for recently reported amplifiers based on SiGe HBT. (b) Gain per stage vs. frequency for the same set of amplifiers.

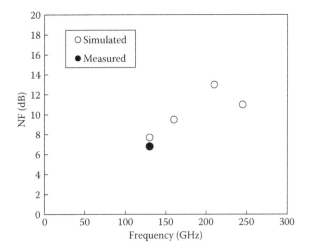

FIGURE 4.11 NF vs. frequency for recently reported amplifiers based on SiGe HBT.

4.3.2 OSCILLATORS

One of the challenges in implementing THz systems is the availability of the high-power signal sources, which critically demand a high-performance oscillator. In addition to such needs in the sources and transmitters to serve as signal generators, oscillators are also an indispensable component for heterodyne systems in both receivers and transmitters that is required to provide local oscillation to mixers. In many cases, the voltage-controlled oscillator (VCO) is a more favored type of oscillator, as the frequency tuning capability is useful for various purposes, including the phase-locked loop (PLL) application as well as the compensation of oscillation frequency shift due to design miss or process variation. Inclusion of varactors for VCO implementation, however, typically calls for more demanding circuit design details due to the low-quality factor of varactors at raised frequency. Methods to improve the quality factor of varactors have also been reported to address such a challenge [28].

Achieving high-frequency operation with oscillators is not straightforward, but it is fair to say it is not as challenging as the amplifier case, the reason being two-fold. First, a gain barely larger than unity suffices to trigger oscillation. This makes oscillators operating near f_{max} of the device feasible. This can be compared with the amplifier case where device f_{max} of at least two to three times the operation frequency is required. Second, the oscillation frequency can be further boosted up by taking its harmonic signals as the output instead of the fundamental signal. The push-push technique is a widely adopted method when a second harmonic of an oscillator is to be extracted, while triple-push oscillators are becoming increasingly popular where a third harmonic needs to be used. In general, the n-push technique can be used to take the nth harmonic from the oscillator, although the output power will be further reduced when higher harmonics are to be taken. For these reasons, oscillators can work at much higher frequencies for a given process technology than amplifiers.

Widely adopted topologies for oscillators based on SiGe HBT technology are *LC* cross-coupled and Colpitts structures. *LC* cross-coupled oscillators are generally known for their simple design and easier start-up condition. With bipolar technologies, though, blocking capacitors are typically needed in the cross-coupling path, along with separate biases for base and collector, which is not necessarily the case for CMOS *LC* cross-coupled oscillators, slightly complicating the design. Colpitts oscillators benefit from the inherent buffering of the *LC* tank from the load [29], which draws wide favor at the mm-wave and higher-frequency bands, although start-up condition is harder to achieve compared to the *LC* cross-coupled architecture.

Reports on Si-based oscillators operating beyond 100 GHz have exploded in recent years, but more of them are based on CMOS, and those based on SiGe technologies are relatively scarce. Figure 4.12 shows the recently reported SiGe HBT oscillators operating beyond 100 GHz, which presents the output power against oscillation frequency. For oscillators operating near 120 GHz, a band that has attracted recent interest because of the industrial scientific and medical (ISM) band located at 122–123 GHz, output power beyond 0 dBm has been reported, reaching up to 6 dBm. At around 200 GHz, reported output power is still lower than 0 dBm, falling in the range of –5 to –10 dBm. The highest oscillation frequency for SiGe oscillators reported so far is around 280 GHz, which was based on Infineon's SiGe HBT technology that yields f_T/f_{max} of 200/275 GHz [30]. The push-push oscillator based on Colpitts topology exhibits an output power of –20 dBm with a rather large DC power dissipation of 132 mW, resulting in efficiency lower than 0.01%. There is a notable trend, though, of significantly improving efficiency, with recent results based on SiGe HBT showing values close to 1% [31]. The best oscillation frequency based on the fundamental mode reported so far with SiGe HBT technology is 181 GHz, which is based on Infineon's SiGe HBT with an f_{max} of 320 GHz [32]. The oscillator, which draws 35 mA from a 1.8 V supply, making up a DC power dissipation of 59 mA (without buffer), is based on Colpitts topology combined with

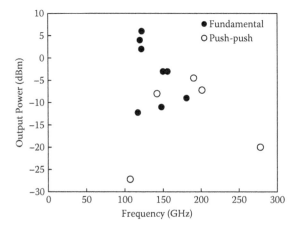

FIGURE 4.12 Output power vs. frequency for recently reported oscillators based on SiGe HBT, shown separately for fundamental oscillators and push-push oscillators.

a capacitive feedback. The phase noise of the oscillator is also reported, which is −82 dBc/Hz. It is noted that it is challenging to measure the phase noise at this range of high frequency. It can be measured at lowered frequency by down-conversion with an external mixer, but uncertainty still remains as the local oscillation (LO) for the mixer might contribute to the total phase noise. For this particular oscillator reported, the phase noise was measured by employing an integrated prescaler, which divides the output frequency by 32, which allows phase noise estimation with an assumption of 6 dB improvement in phase noise for every division by 2.

4.3.3 Mixers

The mixer is a key circuit component in heterodyne systems that down-converts RF signal down to intermediate frequency (IF) band or up-converts IF signal up to RF band. Although mixers can be implanted with passive devices, which benefit from zero power dissipation and better linearity, active mixers can provide a positive conversion gain that helps to suppress the noise from the following stages in addition to adding gain to the system total gain. There have been several reports that detail the characteristics of mixers based on SiGe technology operating beyond 100 GHz. Some attempted to apply Gilbert cell topology, which is perhaps the most popular topology for conventional RF application, to the frequency beyond 100 GHz. While overall acceptable performance was obtained, they typically suffered from excessively high DC power dissipation reaching several tens of mW or higher [33, 34]. Considering the fact that passive mixers are also available free of extra power consumption, such a level of power dissipation for a mixer is a big burden on the total DC power budget for the system that contains the circuit. An alternative approach employing a G_m-boosted technique and a pair of symmetric dual baluns was recently proposed based on TowerJazz 0.18 μm SiGe BiCMOS technology to implement a mixer with low DC power dissipation yet with a positive conversion gain (Figure 4.13) [35].

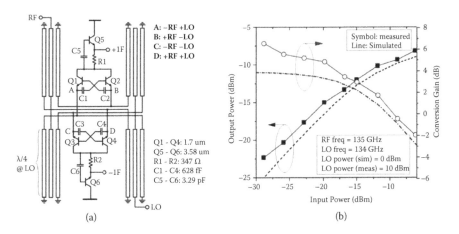

FIGURE 4.13 (a) Schematic of a mixer with G_m-boosted technique and dual balun. (b) Conversion gain and output power of the mixer plotted against input power. (From D. H. Kim and J.-S. Rieh, *IEEE Microwave and Wireless Components Letters*, 22, 409–411, 2012.)

The capacitive cross-coupling acts to boost G_m of the core transistor pair, leading to improved conversion gain that is also helped by the stacked output buffer with current reuse topology employed. The symmetric pair of mixer cores combined with the dual balun provides a differential output with excellent LO-RF isolation. Measured conversion gain is 11.5 dB with an LO driving power of 10 dBm, with a DC power consumption of only 3.9 mW, far lower than those based on the Gilbert cell topology mentioned above.

4.3.4 FREQUENCY DIVIDERS AND MULTIPLIERS

Frequency dividers provide a basic function of dividing frequency by an integer, which is a desired operation in many RF systems. One chief application of the frequency dividers is PLL, where the output frequency needs to be divided down to the reference frequency. There are three types of widely used frequency dividers. Static dividers have been a dominant solution for most applications owing to their wide locking range, but their operation frequency is rather limited and not quite suitable for THz applications. Miller dividers can operate at raised frequency, but only with considerable power dissipation. Injection-locked frequency dividers (ILFDs) can operate at high frequency with low-power dissipation, an attractive feature for THz applications; one drawback is its relatively narrow locking range. While there have been extensive efforts to apply ILFDs for THz by increasing the locking range up to the practical level, they were mostly based on CMOS technology. Most of the reported frequency dividers based on SiGe HBT technologies operating beyond 100 GHz are based on the Miller configuration, except for a few circuits.

One example of ILFDs developed on SiGe HBT is briefly introduced here, which is based on a ring oscillator (RO) [36]. ROs are rarely applied to mm-wave or higher frequency due to their relatively poor phase noise performance at high frequency despite their distinguishing advantage of small size. However, ROs can be favorably applied to ILFDs, since their low Q-factor, which is responsible for the degraded phase noise for oscillator applications, can be actually exploited for ILFDs, as it helps to widen the locking range. The 140 GHz RO-based ILFD shown in Figure 4.14 is based on TowerJazz 0.18 μm SiGe BiCMOS technology, which is composed of a three-stage RO with parallel injection ports. An active load is employed instead of the conventional resistive load for bipolar inverters, which is expected to further increase the locking range. The RO ILFD, which occupies only 0.07×0.03 mm^2 for the core area, shows a locking over the range of 132.5–140.4 GHz with DC power dissipation of 71.2 mW. The size advantage combined with the achieved performance is enough to place RO-based SiGe ILFD as one of the contending candidates for a THz frequency division function.

Frequency multipliers provide an opposite function of dividers, multiplying frequency by an integer. Since raising frequency is a highly desired process for many high-frequency applications, frequency multipliers play a critical role in THz applications such as signal sources, transmitters, and LOs. While a single-frequency multiplier can be sufficient for low-order multiplication, a series connection of frequency multipliers, or a frequency multiplier chain, is often demanded for high-order

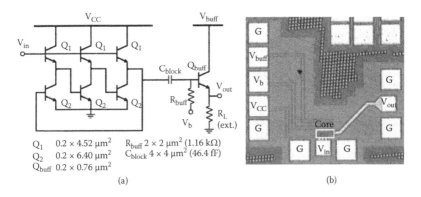

Q_1 0.2 × 4.52 μm² R_{buff} 2 × 2 μm² (1.16 kΩ)
Q_2 0.2 × 6.40 μm² C_{block} 4 × 4 μm² (46.4 fF)
Q_{buff} 0.2 × 0.76 μm²

(a)   (b)

FIGURE 4.14 (a) Schematic and (b) chip photo of a 140 GHz RO-based ILFD based on SiGe HBT. The core occupies only 0.07 × 0.03 mm². (From H. Seo et al., *Electronics Letters*, 48, 847–848, 2012.)

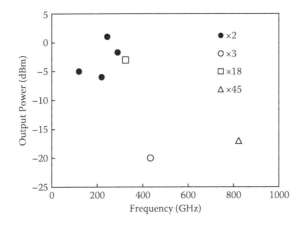

FIGURE 4.15 Output power of multipliers based on SiGe HBT, with various multiplication factors.

multiplication. In this case, it is typical to insert amplifiers in appropriate locations in the chain to recover the power lost by the multipliers, as mentioned earlier. Since there is a strong tendency of output power decreasing with increasing frequency for high-frequency signal sources, the available output power at the output of a single-frequency multiplier or a multiplier chain is probably the principal performance measure of these circuits. Figure 4.15 shows the output power against frequency for recently reported frequency multipliers and multiplier chains. It clearly shows that achieving high power is challenging when frequency is increased. The highest output frequency obtained with a SiGe HBT multiplier chain reported so far is 820 GHz, which is composed of two triplers and one quintupler, together with two amplifiers immersed in the chain [37].

4.3.5 Integrated Transmitters and Receivers

Until the late 2000s the reported Si-based circuits operating beyond 100 GHz were mostly unit circuit blocks. However, as the experiences with such high-frequency bands rapidly accumulate and THz circuit design techniques are maturing fast, the main focus of related research is gradually shifting toward integrated THz circuits. For broadband communication applications, the number of results on integrated THz receivers and transmitters are growing. For imaging applications, integrated THz signal sources and detectors are being reported at an accelerating pace. In fact, the difference between the two types of integrated circuits for the different applications is little. For successful implementation of such integrated circuits, not only correct design for unit circuit blocks but also precise matching between those blocks is highly desired. The challenge for accurate matching becomes more demanding for higher frequencies as the mismatch due to very small unattended dimensional deviations in passive components will result in increasingly more significant design errors. It appears that such risks began to be addressed and reports on THz circuits with high integration levels are growing fast, some of which are introduced below as examples.

Based on a SiGe BiCMOS technology with f_T of 230 GHz from STM, a 380 GHz integrated transceiver chipset was developed by UC Berkeley [38]. The receiver block is composed of a second-order subharmonic mixer, a frequency doubler for quadrature LO generation, and an IF buffer, while the transmitter is composed of a frequency quadrupler with LO driving amplifiers. An LO block, shared by both receiver and transmitter, includes a W-band Colpitts VCO and a branch-line hybrid, implying an overall architecture exploiting the fourth harmonic from the fundamental signal of around 95 GHz. Both receiver and transmitter are terminated with a pair of on-chip patch antennas. The measured effective isotropic radiated power (EIRP) is around –14 dBm to –11 dBm per base bias voltage for the transmitter, while the NF of the receiver is estimated to be 35–38 dB near 380 GHz.

Another noteworthy integrated THz system is the separate receiver and transmitter chipsets operating up to 820 GHz, which were developed by Wuppertal University based on SiGe BiCMOS technology with f_T/f_{max} of 380/435 GHz from IHP [37]. The receiver and transmitter respectively employ a fifth-order subharmonic mixer and 5× frequency multiplier for the frequency down- and up-conversions. The LO is supplied from an external source, which is multiplied up to around 160 GHz for the receiver LO and the input of the last 5× multiplier in the transmitter. Another notable feature of the circuit is that a four-way spatial power combining/division was employed for the transmitter and receiver to improve the output power and sensitivity, respectively. The receiver shows a maximum conversion gain (per channel) of –22 dB and NF of 47 dB, while the transmitter exhibits the best EIRP of –17 dBm.

4.4　CONCLUSION

In this chapter, an overview of the device aspect of SiGe HBTs was presented in terms of their operation principle, DC and RF characteristics, noise performance, and process technology. Furthermore, recently reported performance of SiGe HBT circuits operating beyond 100 GHz was reviewed for amplifiers, mixers, oscillators, frequency divider/multiplier, as well as integrated transmitter/receiver. It is obvious from this review that the development of SiGe HBT technology is still on the right track, fully benefiting from the fact that the costly lithography node evolution is not necessarily needed for HBT performance improvement, unlike the CMOS case. This is a critical advantage for SiGe technology that will be increasingly more appreciated as CMOS technology more heavily depends on an aggressively scaled lateral dimension. Together with the rapidly developing interests in THz applications, the needs for high-performance SiGe HBT technology will keep growing.

REFERENCES

1. S. S. Iyer, G. L. Patton, S. S. Delage, S. Tiwari, and J. M. C. Stork, Silicon-germanium base heterojunction bipolar transistors by molecular beam epitaxy, in *IEDM Technical Digest*, 1987, pp. 874–875.
2. T. I. Kamins, K. Nauka, J. B. Kruger, J. L. Hoyt, C. A. King, D. B. Noble, C. M. Gronet, and J. F. Gibbons, Small-geometry, high-performance, Si-$Si_{1-x}Ge_x$ heterojunction bipolar transistors, *IEEE Electron Device Letters*, 10, 503–505, 1989.
3. E. F. Crabbé, B. S. Meyerson, J. M. C. Stork, and D. L. Harame, Vertical profile optimization of very high frequency epitaxial Si and SiGe-base bipolar transistors, in *IEDM Technical Digest*, 1993, pp. 83–86.
4. D. Ahlgren, M. Gilbert, D. Greenberg, S. J. Jeng, J. Malinowski, D. Nguyen-Ngoc, K. Schonenberg, K. Stein, R. Groves, K. Walter, G. Hueckel, D. Colavito, G. Freeman, D. Sunderland, D. L. Harame, and B. Meyerson, Manufacturability demonstration of an integrated SiGe HBT technology for the analog and wireless marketplace, in *IEDM Technical Digest*, 1996, pp. 859–862.
5. B. Jagannathan, M. Khater, F. Pagette, J.-S. Rieh, D. Angell, H. Chen, J. Florkey, F. Golan, D. R. Greenberg, R. Groves, S. J. Jeng, J. Johnson, E. Mengistu, K. T. Schonenberg, C. M. Schnabel, P. Smith, A. Stricker, D. Ahlgren, G. Freeman, K. Stein, and S. Subbanna, Self-aligned SiGe NPN transistors with 285 GHz f_{max} and 207 GHz f_T in a manufacturable technology, *IEEE Electron Device Letters*, 23, 258–260, 2002.
6. J.-S. Rieh, B. Jagannathan, H. Chen, K. T. Schonenberg, S. J. Jeng, M. Khater, D. Ahlgren, G. Freeman, and S. Subbanna, Performance and design considerations for high speed SiGe HBTs of f_T/f_{max} = 375GHz/210GHz, in *International Conference on Indium Phosphide and Related Materials*, 2003, pp. 374–377.
7. P. Chevalier, F. Pourchon, T. Lacave, G. Avenier, Y. Campidelli, L. Depoyan, G. Troillard, M. Buczko, D. Gloria, D. Celi, C. Gaquiere, and A. Chantre, A conventional double-polysilicon FSA-SEG Si/SiGe:C HBT reaching 400 GHz f_{MAX}, in *IEEE Bipolar/BiCMOS Circuits and Technology Meeting*, 2009, pp. 1–4.
8. B. Heinemann, R. Barth, D. Bolze, J. Drews, G. G. Fischer, A. Fox, O. Fursenko, T. Grabolla, U. Haak, D. Knoll, R. Kurps, M. Lisker, S. Marschmeyer, H. Rücker, D. Schmidt, J. Schmidt, M. A. Schubert, B. Tillack, C. Wipf, D. Wolansky, and Y. Yamamoto, SiGe HBT technology with f_T/f_{max} of 300GHz/500GHz and 2.0 ps CML gate delay, in *IEEE International Electron Devices Meeting*, 2010, pp. 30.5.1–30.5.4.
9. P. Chevalier, T. F. Meister, B. Heinemann, S. Van Huylenbroeck, W. Liebl, A. Fox, A. Sibaja-Hernandez, and A. Chantre, Towards THz SiGe HBTs, in *IEEE Bipolar/BiCMOS Circuits and Technology Meeting*, 2011, pp. 57–65.
10. D. L. Harame, J. H. Comfort, J. D. Cressler, E. F. Crabbe, J. Y. C. Sun, B. S. Meyerson, and T. Tice, Si/SiGe epitaxial-base transistors. I. Materials, physics, and circuits, *IEEE Transactions on Electron Devices*, 42, 455–468, 1995.

11. T. Ishibashi, Nonequilibrium electron transport in HBTs, *IEEE Transactions on Electron Devices*, 48, 2595–2605, 2001.

12. E. O. Johnson, Physical limitations on frequency and power parameters of transistors, *RCA Review*, 26, 163–177, 1965.

13. J.-S. Rieh, B. Jagannathan, D. Greenberg, G. Freeman, and S. Subbanna, A doping concentration-dependent upper limit of the breakdown voltage-cutoff frequency product in Si bipolar transistors, *Solid-State Electronics*, 48, 339–343, 2003.

14. J.-S. Rieh, B. Jagannathan, D. R. Greenberg, M. Meghelli, A. Rylyakov, F. Guarin, Zhijian Yang, D. C. Ahlgren, G. Freeman, P. Cottrell, and D. Harame, SiGe heterojunction bipolar transistors and circuits toward terahertz communication applications, *IEEE Transactions on Microwave Theory and Techniques*, 52, 2390–2408, 2004.

15. Z. Jin, J. D. Cressler, G. Niu, and A. J. Joseph, Impact of geometrical scaling on low-frequency noise in SiGe HBTs, *IEEE Transactions on Electron Devices*, 50, 676–682, 2003.

16. J. A. Babcock, B. Loftin, P. Madhani, X. Chen, A. Pinto, and D. K. Schroder, Comparative low frequency noise analysis of bipolar and MOS transistors using an advanced complementary BiCMOS technology, in *Proceedings of IEEE Conference on Custom Integrated Circuits*, 2001, pp. 385–388.

17. M. Sanden, O. Marinov, M. J. Deen, and M. Ostling, A new model for the low-frequency noise and the noise level variation in polysilicon emitter BJTs, *IEEE Transactions on Electron Devices*, 49, 514–520, 2002.

18. M. J. Deen and E. Simoen, Low-frequency noise in polysilicon-emitter bipolar transistors, *IEE Proceedings of Circuits, Devices and Systems*, 149, 40–50, 2002.

19. J.-S. Rieh, B. Jagannathan, H. Chen, K. T. Schonenberg, D. Angell, A. Chinthakindi, J. Florkey, F. Golan, D. Greenberg, S. J. Jeng, M. Khater, F. Pagette, C. Schnabel, P. Smith, A. Stricker, K. Vaed, R. Volant, D. Ahlgren, G. Freeman, K. Stein, and S. Subbanna, SiGe HBTs with cut-off frequency of 350 GHz, in *IEDM Technical Digest*, 2002, pp. 771–774.

20. B. Geynet, P. Chevalier, B. Vandelle, F. Brossard, N. Zerounian, M. Buczko, D. Gloria, F. Aniel, G. Dambrine, F. Danneville, D. Dutartre, and A. Chantre, SiGe HBTs featuring $f_T > 400$GHz at room temperature, in *IEEE Bipolar/BiCMOS Circuits and Technology Meeting*, 2008, pp. 121–124.

21. J.-S. Rieh, D. Greenberg, A. Stricker, and G. Freeman, Scaling of SiGe heterojunction bipolar transistors, *Proceedings of the IEEE*, 93, 1522–1538, 2005.

22. Y. Shi and G. Niu, Vertical profile design and transit time analysis of nano-scale SiGe HBTs for terahertz f_T, in *Proceedings of the Bipolar/BiCMOS Circuits and Technology Meeting*, 2004, pp. 213–216.

23. Y. Shi and G. Niu, 2-D analysis of device parasitics for 800/1000 GHz f_T/f_{max} SiGe HBT, in *Proceedings of the Bipolar/BiCMOS Circuits and Technology Meeting*, 2005, pp. 252–255.

24. M. Schroter, G. Wedel, B. Heinemann, C. Jungemann, J. Krause, P. Chevalier, and A. Chantre, Physical and electrical performance limits of high-speed SiGeC HBTs. I. Vertical scaling, *IEEE Transactions on Electron Devices*, 58, 3687–3696, 2011.

25. M. Schroter, J. Krause, N. Rinaldi, G. Wedel, B. Heinemann, P. Chevalier, and A. Chantre, Physical and electrical performance limits of high-speed Si GeC HBTs. II. Lateral scaling, *IEEE Transactions on Electron Devices*, 58, 3697–3706, 2011.

26. D. Hou, Y.-Z. Xiong, W.-L. Goh, H. Wei, and M. Madihian, A D-band cascode amplifier with 24.3 dB gain and 7.7 dBm output power in 0.13 mm SiGe BiCMOS technology, *IEEE Microwave and Wireless Components Letters*, 22, 191–193, 2012.

27. Y. Mao, K. Schmalz, J. Borngräber, and J. C. Scheytt, A 245GHz CB LNA in SiGe, in *European Microwave Integrated Circuits (EuMIC)*, 2011, pp. 224–227.

28. Y. Oh and J.-S. Rieh, A comprehensive study of high-Q island-gate varactors (IGVs) for CMOS millimeter-wave applications, *IEEE Transactions on Microwave Theory and Techniques*, 59, 1520–1528, 2011.

29. M. Jahn, H. Knapp, and A. Stelzer, A 122-GHz SiGe-based signal-generation chip employing a fundamental-wave oscillator with capacitive feedback frequency-enhancement, *IEEE Journal of Solid-State Circuits*, 46, 2009–2020, 2011.

30. R. Wanner, R. Lachner, G. R. Olbrich, and P. Russer, A SiGe monolithically integrated 278 GHz push-push oscillator, in *IEEE International Microwave Symposium*, 2007, pp. 333–336.

31. A. Balteanu, I. Sarkas, V. Adinolfi, E. Dacquay, A. Tomkins, D. Celi, P. Chevalier, and S. P. Voinigescu, Characterization of a 400-GHz SiGe HBT technology for low-power D-Band transceiver applications, in *IEEE MTT-S International Microwave Symposium Digest*, 2012, pp. 1–3.

32. M. Jahn, K. Aufinger, T. F. Meister, and A. Stelzer, 125 to 181 GHz fundamental-wave VCO chips in SiGe technology, in *IEEE Radio Frequency Integrated Circuits Symposium*, 2012, pp. 87–90.

33. U. R. Pfeiffer, E. Ojefors, and Z. Yan, A SiGe quadrature transmitter and receiver chipset for emerging high-frequency applications at 160GHz, in *International Solid-State Circuits Conference*, 2010, pp. 416–417.

34. K. Schmalz, W. Winkler, J. Borngräber, W. Debski, B. Heinemann, and J. C. Scheytt, A subharmonic receiver in SiGe technology for 122 GHz sensor applications, *IEEE Journal of Solid-State Circuits*, 45, 1644–1656, 2010.

35. D. H. Kim and J.-S. Rieh, A 135 GHz differential active star mixer in SiGe BiCMOS technology, *IEEE Microwave and Wireless Components Letters*, 22, 409–411, 2012.

36. H. Seo, J. Yun, and J.-S. Rieh, SiGe 140 GHz ring-oscillator-based injection-locked frequency divider, *Electronics Letters*, 48, 847–848, 2012.

37. E. Ojefors, J. Grzyb, Y. Zhao, B. Heinemann, B. Tillack, and U. R. Pfeiffer, A 820GHz SiGe chipset for terahertz active imaging applications, in *IEEE International Solid-State Circuits Conference*, 2011, pp. 224–226.

38. J.-D. Park, S. Kang, and A. M. Niknejad, A 0.38 THz fully integrated transceiver utilizing a quadrature push-push harmonic circuitry in SiGe BiCMOS, *IEEE Journal of Solid-State Circuits*, 47, 2344–2354, 2012.

5 Multiwavelength Sub-THz Sensor Array with Integrated Lock-In Amplifier and Signal Processing in 90 nm CMOS Technology

Péter Földesy

CONTENTS

5.1 INTRODUCTION

The architecture and the operation of a sub-THz sensor array are presented, as implemented in standard 90 nm complementary metal-oxide-semiconductor (CMOS) technology. The integrated sensor array is arranged around 12 silicon field effect plasma wave detectors with integrated planar antennas. The received signals are further processed by preamplifiers, analog-to-digital converters, and a time-shared digital domain lock-in amplifier. The system automatically locks to external modulation and provides standard digital streaming output. Instead of building a uniform array, seven different antenna types with various polarization properties (horizontal and vertical linear, left- and right-handed circular polarization) and spectral responsivity have been integrated. The sensors altogether provide broadband response from 0.25 to 0.75 THz. The peak-amplified responsivity of the sensors is 185 kV/W @ 365 GHz, and at the detectivity maximum the noise equivalent power (NEP) is near 40 pW/$\sqrt{Hz}$. Relying on the drain current-induced responsivity increment, this peak value rises above 1.2 MV/W with a moderate NEP ~200 pW/$\sqrt{Hz}$ at 50 nA source-drain current. Two application examples are provided as well: a multiwavelength transmission imaging and a homodyne imaging case with complex amplitude recording.

The THz spectrum of electromagnetic waves is nonionizing and has a broad application area [1]. In [2] Dyakonov and Shur predicted that the instability of electron plasma waves in short-channel field effect transistors (FETs) could be used as a terahertz frequency radiation detector. A different, more phenomenological description is given in [16] based on resistive self-mixing. Several THz imaging systems and different sensor technologies appeared, among others, silicon-based field effect transistors [2–6]. Besides high-mobility, it found that not only high-mobility devices, but also silicon-based detectors with integrated planar antennas, could serve as fast imagers as well [9, 10, 15–22]. The silicon- or SiGe-based sensor technologies offer an advantage over other material-based solutions, like bolometers, the on-chip integration of readout and signal processing circuitry [15]. Related to FET detectors, we can distinguish two basic operation modes: open drain and nonzero drain current cases. The former provides higher sensitivity, while the latter provides a significantly higher response [7, 8], though with dominant flicker noise. In homodyne and heterodyne mixing FET detectors are presented [4, 12] with outstanding performance, similar to what Schottky diodes offer. These results suggest that still further improvements are expected from silicon-based sensors.

As the photodetectors advanced from a few passive pixels to single-chip video cameras, the same level of integration can be easily imagined for THz range imagers as well. This work is a step toward integrated imagers, by the inclusion of digitalization, digital postprocessing, and output streaming along the high-sensitivity and versatile sensors.

5.2 SYSTEM ARCHITECTURE

The presented sensor array contains 12 antenna-detector pairs and the following analog and digital circuitry. The system has been designed and manufactured using standard 90 nm Taiwan Semiconductor Manufacturing Company (TSMC) technology. The motivation behind the system development is to investigate various antenna configurations operational from 0.25 to 0.75 THz and create an integrated smart pixel array for autonomous image acquisition.

From an engineering point of view, during the design with FET plasma wave sensors some fundamental difficulties should be addressed. One of the issues comes from the nature of the FET detectors [6]: the output signals (potential difference generated between the source and drain terminals) of the sensors are small ($\mu V - mV$). The typically subthreshold operation of the FETs provides low driving capability with output resistance of $k\Omega - M\Omega$. Hence, there is also a need for signal amplification with low input capacitance near the detector. This high output impedance results in relatively low-frequency operation into a region where the $1/f$ noise becomes significant. The detector's major noise contribution in open drain mode originated in its thermal noise across the channel, of which the power spectral density varies from a few dozen $nV/\sqrt{Hz}$ to $\mu V/\sqrt{Hz}$ for typical detector solutions. In nonzero drain current mode, the additional flicker noise further increases this value [7]. Though the sensor noise may rise to high values, the signal amplification must not add excess noise to the detector signal. These properties pose a practical limit on the amplifier design. The next problem to be solved is the free space coupling of the radiation to the detectors as the coupling determines the final performance. In standard silicon technologies the doped substrate constitutes a high loss factor and the metallization (number of metal layers, thickness, dielectric) is predetermined. As a result, resonant structures with ground shielding are relatively straightforward to design [34], while high-sensitivity broadband receivers are more difficult without microelectronic postprocessing (e.g., cavity etching under the antenna structure [17]). Finally, the high-impedance detector signals are susceptible to digital noise coming from other parts of the system, which requires careful mixed-signal design practice.

5.2.1 FUNCTIONAL STRUCTURE

The chip comprises 12 sensors arranged in a 4-by-3 array [21, 22]. The sensors are composed of an antenna-coupled FET detector followed by low-noise amplification. Next, each amplified sensor signal is digitalized and demodulated by a digital lock-in amplifier. The FET detectors are identical in each channel, while the antenna structures are different, including spiral, bow tie, and dipole antennas. For signal amplification an AC-coupled single-ended operational amplifier has been

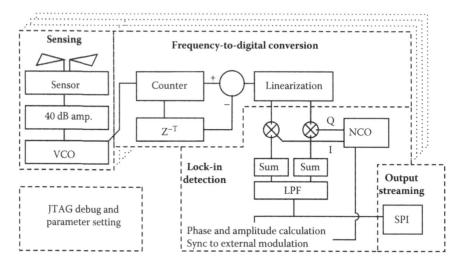

FIGURE 5.1 System architecture. Components are antennas, detectors, low-noise amplifiers, VCOs and followed by frequency estimation, linearization modules, and common time-sharing lock-in detection comprising low-pass frequency filtering, numerically controlled oscillator (NCO), and output streaming.

integrated for each sensor. The digitalization is achieved by a voltage-controlled oscillator (VCO) and frequency estimation pair with calibration. The digitalization is followed by channel-wise lock-in detection. The implemented lock-in detection is based on amplitude modulation of the irradiation (or sensitivity modulation of the detectors) and complex-valued demodulation of the amplified sensor signal. There are two modes available: when the system provides modulation for the external radiation source and a complementary mode, and when it can synchronize to an external modulation coming from, e.g., a mechanical chopper. The demodulated responses of the channels are then low-pass filtered and selectively sent over a serial peripheral interface (SPI) port. The debug capabilities of the system are broad and reachable through a joint test action group (JTAG) interface. The conversion parameters, the modulation frequency, and the low-pass filter parameters can be set via this interface. The reason for embedding a JTAG control interface is to reduce the pin count of the chip and to provide a simple and structured way to access its hundreds of control bits and internal states. The system functional architecture can be seen in Figure 5.1. The details of these functionalities and physical architectures are described in the following.

5.2.2 OPTICAL CONSIDERATIONS

One could involve another constraint in the sensor array design as well: the optical properties of the system as an imager. In the presented system, instead of pursuing classic imager style operation, the focus has been moved to multiwavelength operation with polarization variants. The reasoning behind this follows.

On tabletops and short distances the THz imagers, including pulsed photo-conducting architectures, usually mechanical scanners, move the field of view or the object to collect the information from pixel to pixel or for a small pixel array. Most of the imaging platforms are built around reflective components in order to shape broad spectral range equally (e.g., broadband sources radiate in the 0.3–3 THz range). As a consequence of reflective optics, with a few additional refractive elements, the achievable focusing capability is restricted. The reported numerical apertures vary from 0.4 to 0.05, resulting in a 2 to 10 times larger spot size than the actual wavelength. Due to the resonant antenna size reduction implemented in a substrate, the optimal antennas are typically much smaller than what the optics can resolve. In other words, the effective area of a single antenna with matched resonant peak is much smaller than the reasonable spot size. Taking a practical example of a focal plane detector on silicon substrate, the difference between the optical resolution and the antenna size could be an order of magnitude. Using an $f/1.4$ optics and operational frequency at the water absorbance peak at 0.55 THz ($\lambda = 0.55$ mm), the optimal dipole antenna length would be 108 µm [36], while the spot size full width at half maximum (FWHM) diameter becomes about 1 mm ($D = 1.22\lambda_0 f_{\#}$).

One can find solid immersion lens solutions in a form of silicon elliptical or hemispherical lenses [24–26]. On the other hand, these solutions have a substantial decrease in amplitude at high frequencies and strong frequency dependence [23]. Another way is to integrate many individual sensors in a multichip module [25] to mitigate the large difference between antenna size and free space wavelength.

5.2.3 PHYSICAL ARCHITECTURE

The physical floorplan is determined by the sensor array. The antennas require definite area, while the digital and analog circuitry could be placed practically in any shape and aspect ratio. First, the targeted frequency range is selected (0.25–0.75 THz) and the main structural dimensions of the antennas are calculated. Seven different antenna structures are designed with different polarities based on knowledge of the technological fundamentals, such as metal layers, their thickness, and dielectric properties. Next, using the antenna dimensions, mutual crosstalk requirements, and distance from the package wire bonding, the unit cell is fixed and the available area is divided into 12 such cells of equal size with a pitch of 330 µm. The digital circuitry takes place on a distant corner and the analog front-end circuitry is integrated within the sensors with careful electromagnetic (EM) shielding. The resulting floorplan is presented in Figure 5.2.

In order to make the design phase more efficient, each cell has exactly the same circuitry and layout, except for the antennas. This way, antennas of different kinds are placed in a similar metallic environment and the circuit design and verification are simplified greatly. The cell layouts are placed rotated and mirrored to share power distribution lines and to save area by overlapping metal-insulator-metal (MIM) capacitor bounding structures. The concept can be seen in Figure 5.3.

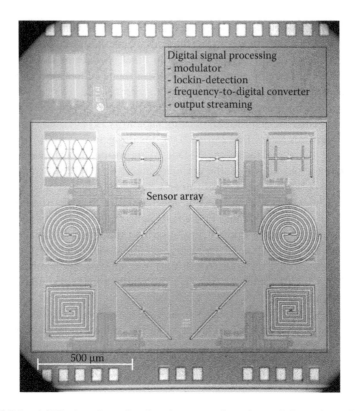

FIGURE 5.2   ASIC microphoto showing the system floorplan and the variety of antenna-coupled sensors.

A constraint must be taken into account; namely, the commercial CMOS technologies are built on conductive, lightly doped silicon substrate. The antennas directly implemented on such a substrate suffer from high substrate losses. A trivial solution is to create a metal shielding beneath the antennas and right on top of the substrate. Thus, in order to avoid high substrate losses, each antenna is placed on the top metal layer with a ground metal mirror underneath in the lower metal layers. One exception to this rule is the sensor of the bow tie antennas. The reason is that the other types are resonating at specific frequencies, while the bow tie antennas are designed to be broadband. Though it responds as expected, its sensitivity is below the resonating and shielded ones, and as shown later, the response is significantly altered from the theoretical one due to substrate resonances. More precisely, as Figure 5.3 illustrates, the substrate is covered by a ground shield formed by the two lowest metal layers (MET1–2). The reason for the double layer is that the signal paths reaching the detectors in the middle of the cells are also covered this way: they run in the lowest metal, while the second metal remains intact. The next structural element is another shield layer created on a higher metal layer, namely, in the sixth one (MET6). This shield covers the majority of the circuitry and is placed at the border of the cells. This layer is also used for power distribution.

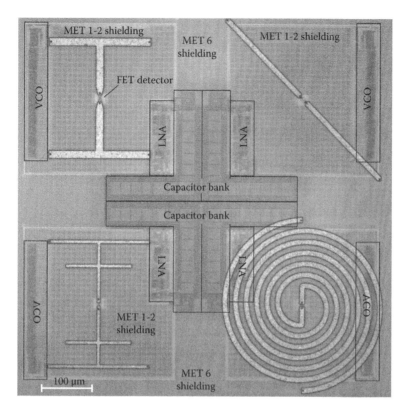

FIGURE 5.3   Floorplan of four adjacent sensor cells with their main components.

At last, another role of this shield is to embed the selection and other signals coming or going to the digital region in the intersensor channels. The antennas are formed on the top metal layer. This particular technology offers nine metal layers for routing, with thicker ones on the top, and an additional thick layer, called the power distribution layer, of 3 µm above the routing layers. From an EM design point of view, the height of the antenna structure above the ground shield is advantageous to increase; hence, the top layer is picked for the antennas. Near the FET detectors, the lowest metal shields are opened and a staggered structure has been built from lower metals and vias. The exact structure affects the behavior of the antennas and needs detailed EM simulation to maximize the electric field concentrated on the detector FET. The illustration of the role of the different layers is shown in Figure 5.4.

5.3   FREE SPACE COUPLING

The FET detector sensitivity and operation are determined by the manufacturing technology and materials (e.g., minimal feature size, silicon or compound material, etc.) and do not alter significantly across the same feature-sized CMOS

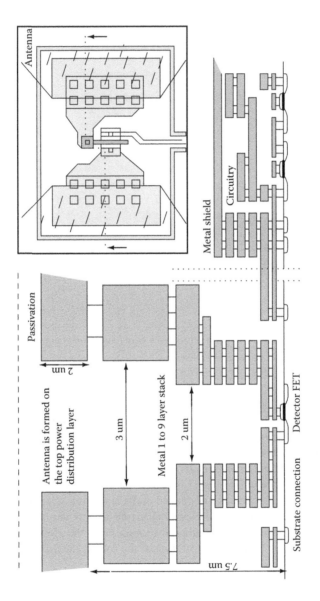

FIGURE 5.4   The metallization strategy near the detector FETs and the nearby circuitry shielding can be seen. The upper right inset shows the bird's-eye view of the detector connection layout to the antenna wings.

technologies [10, 12, 18]. On the other hand, what crucially affects the overall sensitivity is the antenna design and coupling strategy. The basic element of radio frequency (RF) microelectronic technologies is the integrated planar antenna. In the design of an integrated antenna one can choose broadband (e.g., bow tie) or narrow band (e.g., patch) antenna types from the numerous implementation styles [33–35]. The implemented antenna variants are easy to identify on the microphoto in Figure 5.2 and are the following:

- Dipole of different polarization directions (resonate at 0.25 THz)
- H-shaped dipoles (resonate at 0.36 and 0.45 THz)
- Two-armed Archimedean and squared spirals (with 8 and 10 significant resonant peaks from 0.15 to 0.5 THz, respectively)
- Serially connected bow tie antennas on silicon substrate (with peaks at 0.44, 0.55, and 0.7 THz)

The antenna design follows classic methodology. First, the spectral response is selected and the type of antenna (bow tie, spiral, dipole, H-shaped dipole). Next, based on the technology parameters, a quick, textbook estimate is calculated [35]. The responsivity peak(s) of the antenna could be estimated by a simple quasi-static approach up to 1.5 THz according to [36]. At last, an EM field simulator is used to find out exact dimensions, still using ideal metals and a dielectric structure. The actual layout further changes the ideal EM simulation result, caused by slotting design rules and the quantized shape of the vias. This alteration can be verified and fed back to the sizing by exporting the layout from the CAD tool and importing it into the EM simulator. As a last step, the design of the detector coupling is optimized from an electromagnetic point of view with the EM field simulator and simplified plasma wave detection modeling [9]. As a distinguishing step from the classic antenna design, the electric field strength is maximized on the FET terminals, instead of relying on the S-parameters.

5.4 MIXED-SIGNAL INTERFACE

The analog interface design covers the analog front end and the digitalization. In the front end, the requirements are derived primary from the FET detector behavior: the high output impedance, band-limited modulation frequency, and near-ground small signal values. The choice was an AC-coupled single-ended amplifier with low cutoff frequency and small input capacitance. As an additional feature, a simple way is provided to characterize a nonzero drain current configuration. The circuitry details are described next.

5.4.1 ANALOG FRONT END

The FET detector sensitivity increases, lowering the gate-substrate voltage due to the electron plasma thinning under the gate electrode [5]. The downside of this sensitivity increase is that as the transistor penetrates deep into subthreshold, the output signal becomes vulnerable to thermal noise and the FET's output impedance rises exponentially. The basic solution is to terminate its drain with a high-value resistor toward ground level and measure the potential appearing on this resistor, or to inject nonzero DC source-drain current and measure the voltage drop on the detector. In the latter mode, the responsivity of the sensor significantly increases, though the sensor flicker noise increases as well at almost the same rate as presented in [7] in case of high electron mobility transistors. In order to get rid of broadband noise, usually radiation modulation is applied, along with a lock-in amplifier. This way the $1/f$ noise can be efficiently suppressed as well. Hence, the analog signal conditioning must have low input capacitance (~pF) to allow adequate modulation frequency to avoid low-frequency noise. The open drain detectors produce near-ground-level signals, so capacitive coupling had been chosen to get rid of the low DC signal.

The signal amplifier is motivated by neurobiological solutions, due to the similarity of low signal levels and low-frequency modulation [27, 28]. The implemented circuit provides 40 dB amplification with 250 Hz to 0.5 MHz lower and upper –3 dB cutoff frequencies. The amplifier has a folded cascaded structure with a p-channel metal-oxide-semiconductor (PMOS) field effect transistor (Figure 5.5).

It has a small footprint and moderate noise contribution (~40 nV/√Hz at 1 kHz input referred noise), which is smaller than the FET detector thermal noise in its high-sensitivity region. In order to minimize the substrate noise, the detectors and the amplifier are protected by deep n-well guard rings. The band pass filter is based on the high-resistance pseudoresistor and metal-oxide-metal capacitor. The higher frequency limit comes from the lock-in-oriented operation, i.e., physical light chopper and radiation source modulation frequency. The lower cutoff frequency is intended to suppress the increasing $1/f$ noise of the amplifier. Note that the operational conditions of the FET detector at its noise equivalent power (NEP) minimum also limit the upper modulation frequency to about a few kHz. The band pass amplifier topology can be seen in Figures 5.6 and 5.7 for the single and multiple detector configurations.

The FET detector can be considered part of the analog front end. The detectors are identical in each pixel and have drawn a size of W/L = 500 nm/100 nm. The detector transistors are connected so that the gate is fed by an antenna wing; their drain is connected to the other antenna wing and the DC shortened to ground, while the source side feeds the amplifier. This particular connectivity produces negative source-side signals [17], which are inverted by the amplifier topology. In order to support a nonzero drain current configuration, a PMOS-based pseudoresistance of high nonlinear resistivity is integrated. The pseudoresistance behaves as a linear resistor (in the range of GΩ to TΩ) at low potential difference between its terminals [27]. On the other hand, its conductance increases significantly by increasing its terminal voltage (>0.7 V). Connecting to the FET detector, it allows driving 0–5 μAmp source-drain current, which can be set and monitored outside. Until the response of the detector remains in the mV range, the linearity, high value, and fine tunability of the resistance are maintained. The great advantage of this solution is that it has a small footprint, and has very low noise compared to any current source-based architecture.

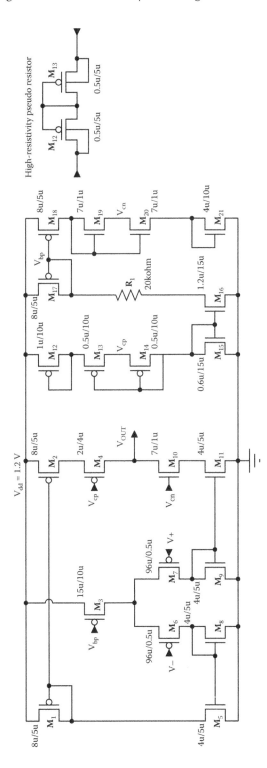

FIGURE 5.5   On the right side, the amplifier with biasing network is shown. In the upper left corner, the implemented pseudoresistor can be seen.

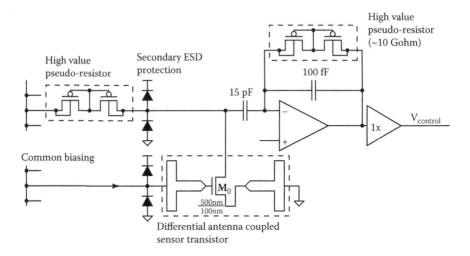

FIGURE 5.6 Simplified schematic of the H-shaped dipole antenna-coupled detector and the band pass amplifier.

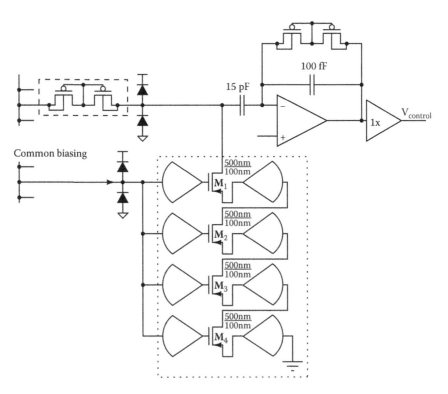

FIGURE 5.7 Simplified schematic of the serially connected multiple bow tie detectors and the amplifier.

Note the electrostatic discharge (ESD) protection on the detector ports as well. First, the reason for including such secondary ESD protection is that the gate signal is provided externally and the pad ESD circuitry is routed far from the sensors. Second, the large antenna area may alter the transistor gate threshold level during manufacturing in an unpredictable way (so-called antenna rules are met this way).

5.4.2 DIGITALIZATION

A free-running VCO and a frequency estimator generate digital representation of the amplified sensor response. The reason for this choice is the limited layout area and metallization that restricts the ADC complexity, and the voltage-to-frequency converter has a small footprint with reasonable precision (25 × 230 µm). The VCO is based on a five-stage interpolating voltage-controlled delay line, which provides low jitter [29]. In order to reduce the clock feedthrough from the oscillator to the sensor via the capacitive amplifier, distributed source followers and a unity buffer are inserted. The VCO can be tuned by the variable capacitive load that supports a wide frequency range (180 ~ 260 MHz) centered at 230 MHz at the nominal half-supply voltage of 0.6 V bias and ±0.3 V control voltage swing. The VCO architecture is shown in Figure 5.8 and a single stage of the ring oscillator is presented in Figure 5.9.

The frequency of the VCO is first divided by two at the VCO to obtain 50% duty cycle and measured by a counter, whose increment is sampled by regular time periods. In addition to the static nonlinearity, differences arise in the oscillation frequency of the VCOs due to manufacturing inaccuracies. These effects are compensated after digitalization by a second-order polynomial equation. This plain ring oscillator is sensitive to temperature and power supply changes as well. It is placed on a separate power domain, which is supplied by an external LDO with high-voltage stability.

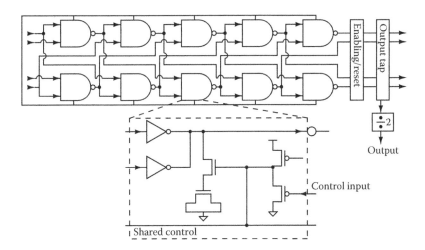

FIGURE 5.8 The implemented low-jitter voltage-controlled delay line with kickback-reduced control input. Its output is fed back to its input to form a voltage-controlled oscillator.

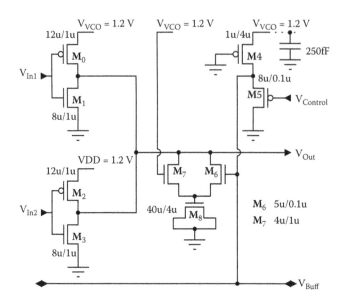

FIGURE 5.9   One stage of the voltage-controlled delay line.

The temperature dependence is handled by repeated measurement cycles using the test access points before actual measurements, and its results are incorporated from time to time into the compensation formula. The corrected 24-bit fixed-point output values (8-bit fractional) are sent toward the lock-in detection. Under nominal operating conditions, the small signal conversion factor of the VCO is 125 MHz/V with a tracking jitter near 350 ps. Though it is a high value for jitter, taking the practical timing conditions of pixel acquisition into account, the overall performance does not deteriorate the results significantly. The VCO uncertainty at a 1-KHz conversion rate yields approximately 0.03 MHz frequency estimation standard deviance. Supposing 1 and 10 pixels per second acquisition rates, it would correspond to 70 and 230 $nV/\sqrt{Hz}$ detector-referred noise.

5.5   DIGITAL CIRCUITRY

The digital circuitry is responsible for handling the streaming data coming from the 12 sensors and maintains communication with the external host. The reason for the inclusion of such functionality into the system is straightforward: make the operation simple and versatile. The implementation style was standard digital flow based on manufacturability, testability, and low-power standards.

5.5.1   CONTROL AND MONITOR

The check and control of the digital part of the system is twofold: first, its correct operation is verified; second, its operational parameters are set. In order to balance the chip area overhead and testability, only specified points are monitored and substituted in the data flow. The test development, including test generation and fault simulation and coverage, has been conducted. The fault-free response is checkable by binary vectors at the monitored buses and cyclic redundancy check (CRC) checksums. The settable parameters and the test points are chained in different domains of a standard JTAG controller, and the external pads are chained in a standard boundary scan [37, 38]. The JTAG interface-handled chains are the following:

- Analog block (including low-noise amplifiers (LNAs)) enabling register
- Analog/digital interface enabling register (including VCO control)
- Digital core configuration register (various parameters, such as modulation frequency, mode of locking, oversampling of the digitalization stage)
- Monitoring and substitution of the digitalization stages
- Analog access port control register
- Digital core, various output sample registers

5.5.2 LOCK-IN DETECTION

As usual in low-signal-level situations, the sensor responses are detected by the lock-in technique. This method increases accuracy and precision by acquiring information with modulated stimulus and integrating the samples over a large time period with low-pass filtering. Correspondingly, the core of the digital part is a digital lock-in amplifier. It consists of in-phase and quadrature phase demodulation, low-pass filtering, and data streaming modules (Figure 5.1). The digital lock-in detection is capable of locking to the source modulation frequency and provides the complex and absolute intensity of the input signal at the selected frequency bin. Besides this mode, the chip can generate a modulation signal as well.

The system performs signal amplification and digitalization per sensor channel in a distributed way, while the lock-in detection is done by a time-shared data path arithmetic logic unit (ALU). The ALU is a fixed-point integer with changing fractional and integer bit length. This module processes all sensor channels or a specific physically meaningful selection of them, while the nonused sensors are usually switched off by the JTAG interface. The simple formula of the complex-valued calculation in a single bin is as follows:

$$Y_B^W = I_B^W + jQ_B^W = \sum_{n-0}^{W-1} X_{\text{sensor}}(n) * \exp^{\frac{i*2\pi}{W}*B*n}$$

where W and B denote window size and frequency bin, respectively. The real and imaginary parts (I, Q) are processed independently. The calculation of the single sensor is done by summing the product of the input time series with sin/cos waveforms representing the in-phase and quadrature phase. The waveforms are generated from a lookup table, which is part of a numerically controlled oscillator (NCO). This NCO derives a fractional frequency from the main system clock and periodically provides the sin/cos values. The intensity value of a specific sensor y_{ch} is calculated then as

$$y_{sensor} = Y(\text{sensor}) = \frac{\sqrt{I_{sensor}^2 + Q_{sensor}^2}}{W}$$

The numeric representation of the sin/cos waveform is signed 12-bit with 11-bit fractional, the I, Q values have 36-bit with 11-bit fractional, and the output is 40-bit with 8-bit fractional. The final output data consist of the real and imaginary parts and the intensity value of the selected frequency bin. When a complete time window is ready, all results of the enabled channels are streamed out by a standard SPI bus.

5.6   TESTABILITY FEATURES

Besides the normal operation of the sensor array, there are numerous test requirements. Such requirements range from near-thermal noise measurement of the sensors under sub-THz irradiation to multichannel lock-in detection data path verifications. The system is prepared for these cases by JTAG-controlled fine-grade on/off power domains, digital access ports, and standard mixed-signal test solutions.

First, multiple power domains have been implemented. The reason is primarily to support clean power supply for the analog front ends and decouple the VCO high-frequency noise from the rest of the circuitry. The secondary motivation is to provide a noise-free environment for the FET detector characterization by completely switching off the VCO and digital regions. Though the antenna-coupled plasma wave detector behavior could be modeled to a certain complexity, the RF characterization of the sensors and the following low frequency (LF) signal path must be separate. The mixed-signal path verification and characterization is available by analog boundary modules [38]. These modules are placed in order to facilitate analog parametric characterization. On the other hand, if the FET detector alone is under test, it is useful to detach, for example, its noise performance from the rest of the circuitry. There are a set of digitally controllable analog boundary cells implemented near the sensors: amplifier and VCO input and outputs can be selected for tests and disconnected from the signal path. Their signals are connected to analog test I/O pins. In a similar way, external input voltage is provided for the VCOs, whose outputs are directly accessible at an I/O pin as well.

5.7   RF CHARACTERIZATION

The radiation source used is a YIG oscillator-driven VDI amplifier/multiplier chain (AMC) to generate an 80 to 750 GHz signal. The polarized signal is radiated through a horn antenna and collimated and then focused by off-axis parabolic mirrors. The responsivity has been calculated by raster scanning the spot size and scaling the beam power measured by a VDI Erickson absolute power meter to the integrated response [17]. The different antenna structures provide various frequency responses. The resonant peaks are drifted in frequency and spread over the resonant peaks compared to the EM simulations. The following data are listed for the highest responsivity structures: the narrow band H-shaped dipole and broadband four serially connected bow ties [30, 31]. The detector noise characterization is performed under a battery-powered setup with switched-off VCO, digital signal processing domains, and using external low data acquisition.

5.7.1 RESPONSIVITY

In the plasma wave sensor context, the RF characterization typically embeds a frequency-dependent response in conjunction with detector biasing. The biasing means gate-source voltage (VGS) and drain-source current (IDS) sweeps of the sensor transistor. Another important feature of such sub-THz sensors is the noise equivalent power, whose measurement requires a noiseless environment. The RF setup based on a sub-THz source is the ensemble of a YIG oscillator and a multiplier amplifier chain operating from 80 to 750 GHz. The irradiation power reaching the sensor plane is validated by a bolometric absolute power meter through a high-resistivity silicon (HRFZ-Si) beam splitter. The focusing is helped by a fiber-coupled visible diode laser and an indium tin oxide (ITO)-covered glass mirror (see Figure 5.17). The quasi-optical setup enables automated modulation frequency, polarization, and attenuation control. These features are generated and the detector response is captured by a standard data acquisition system.

The responsivity was measured according to [17]. This technique is based on raster scanning a focused spot and normalizing the received response with the beam power and the effective area of the detector:

$$R_{det} = \frac{U_{det}}{P_{det}} = \frac{\iint_{area} U_{det}(x, y)dx\,dy}{P_{beam} * A_{det}}$$

The responsivity can be calculated, where U_{det} is the DC photoresponse, P_{beam} is the total beam power in the detector plane measured with a large-aperture THz bolometer power meter, and A_{det} is the detector size. In [17] A_{det} is chosen to be the physical size of the detector, i.e., the pitch size. It is well known that an antenna receives the incident power density by its effective area. The effective area usually does not equal the physical area. In [16] the same scanning process was performed but interpreted in a different way, which is adopted here. Namely, they simulated the antenna's directivity and applied the following well-known equation:

$$A_{eff} = D\frac{\lambda_0^2}{4\pi}$$

where D is the directivity of the on-chip planar antenna and λ_0 is the free space wavelength. However, it must be emphasized that this expression is valid if and only if the antenna is used under the following conditions: (1) lossless and (2) impedance and polarization matched over the whole frequency band, where is it functioning.

Based on the formulated responsivity measurement, several aspects of the sensors can be characterized. As described in detail in [11], the loading effect is presented first. The responsivity is measured as a function of the source-gate potential and the radiation modulation. By these two parameters the maximal response can be found for a system, which is limited by the detector and the following resistive/capacitive loads, such as amplifier input capacitance and bandwidth. Figure 5.10 presents the results of the H-shaped antenna-coupled sensor behavior as a function of gate-source voltage and the input radiation source modulation frequency. Its frequency response is limited at low frequencies by the amplifier roll-off and at higher frequencies by the radiation source used. Nevertheless, the resulting curves suggest that the peak responsivity is near 190 kV/W at 1–2 KHz modulation frequency.

Next, the nonzero source-drain current effect is measured. The drain current increases the response of the detector [7] with the price of increasing $1/f$ noise. Figure 5.11 is presented as a reference for the source-drain current dependence on V_{PR}. Fixing the modulation frequency at 1 KHz, the source-drain current effect is measured and shown in Figure 5.12. The resulting curves are similar to the expectations, with the difference that the enlargement of sensitivity is lower than that found in the high electron mobility transistor (HEMT) case. In the case of this antenna, the largest responsivity values reached 1.2 MV/W. The drain current was injected by applying external voltage on the drain-side pseudoresistance. Its resistance decreases significantly, and lets current flow through the detector FET. The current is measured externally.

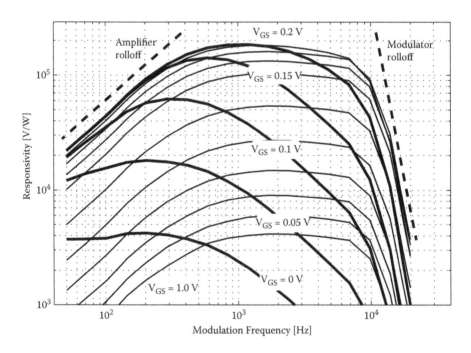

FIGURE 5.10 The H-shaped dipole antenna-coupled sensor responsivity at 360 GHz as a function of gate-source voltage and the input radiation source modulation frequency.

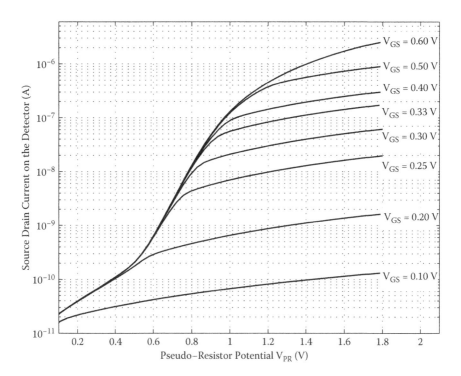

FIGURE 5.11   Source-drain current versus the gate-source voltage and the pseudoresistance potential.

5.7.2 Noise

The next important question is the noise performance. In an open drain configuration, the noise is dominated by thermal noise of the detector channel as $N_{det} = 4kT/G_{DS}$, where G_{DS} is the channel conductance at zero drain-source voltage. Additional noise sources such as noise due to the distributed substrate resistance and shot noise associated with the leakage current of the drain-source reverse diodes are neglected in this calculation. If current flows through the detector, the additional flicker noise appears, which is proportional to $1/f$ as $N_f = K_f/(C_{ox}^2, W, L, f)$, where W, L, C_{ox} are parameters of the transistor, and K_f is a technology-dependent constant [28]. Finally, the amplifier input-related noise is taken into consideration as $N_A(f)$. The noise equivalent power (NEP) is an important figure of merit that describes the minimum power detectable per square root of bandwidth, which is defined as

$$NEP_{system} = \frac{Total\ noise\left(\dfrac{V}{\sqrt{Hz}}\right)}{Responsivity\left(\dfrac{V}{W}\right)} = \frac{\sqrt{N_{det} + N_f + N_A + N_{ADC}}}{R_{det}}\left(\frac{W}{\sqrt{Hz}}\right)$$

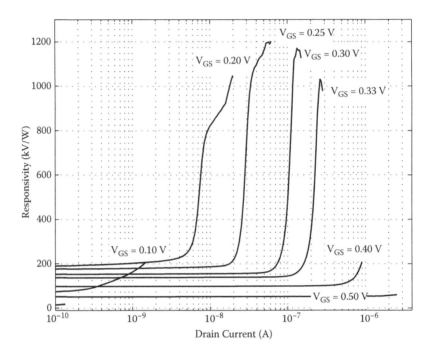

FIGURE 5.12 The H-shaped dipole antenna-coupled sensor's amplifier responsivity as a function of source-drain current injected through the pseudoresistance.

Figure 5.13 shows the NEP for different source-gate voltages as a function of the pseudoresistor potential (V_{PR}). What is important to note is that the NEP does not improve significantly at any point as the source-drain current increases, as found in the HEMTs. On the other hand, at $V_{GS} \sim 0.25V$, $V_{PR} \sim 1.6V$, and $I_{DS} \sim 50nA$, the responsivity increases to the *MV/W* region with a moderately increased NEP value of about 200 pW/$\sqrt{Hz}$.

5.7.3 Design Examples

In the following, two particular sensors are described in detail: the bow tie and the H-shaped dipole antenna. The response and the design metrology are different for these basic types. The microphotos with the physical dimensions are shown in Figure 5.14.

5.7.3.1 Resonant Antenna

In general, the effective permittivity (ε_{eff}) of the substrate-air half spheres tunes the fundamental resonant length (L_R) of the antenna below compared to their free space counterparts (L_0):

$$\varepsilon_{eff} = \frac{\varepsilon_0 + \varepsilon_{substrate}}{2}, \quad LR = \frac{L_0}{\sqrt{\varepsilon_{eff}}} \, \alpha \, \frac{\lambda_0}{\sqrt{\varepsilon_{eff}}}$$

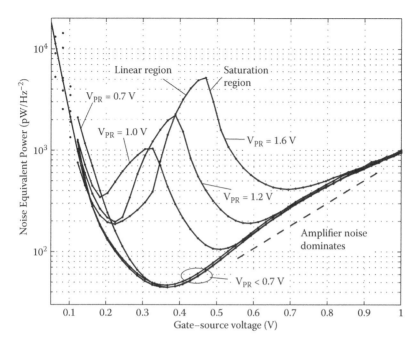

FIGURE 5.13 The H-shaped dipole antenna-coupled sensor's noise equivalent power at 360 GHz as a function of gate-source voltage and potential applied on the pseudoresistor load.

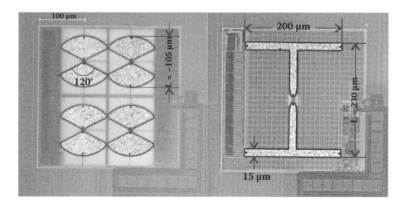

FIGURE 5.14 Microphotos of two specific sensors with broadband and narrow band responses: with bow tie and H-shaped dipole antennas.

The resonant length of an antenna implemented on silicon substrate ($\varepsilon_{Si} = 11.69$) becomes 2.5 times smaller this way. In the mainstream technologies, the metallization is embedded into a sandwiched dielectric structure (combination of SiO_2 or low-K materials with dielectric constant $\varepsilon \sim 2, \ldots, 3.9$) and covered on the top as well (e.g., with Si_3N_4 with dielectric constant $\varepsilon \sim 7.5$). This dielectric coverage, though as thin as 10 μm, also alters the ideal antenna behavior and shifts down the resonance frequency.

The H-shaped antenna resembles the photoconductive antenna designs usually found in pulsed THz systems. It can be considered a dipole of the length L, with a correction of the effective dielectric constant calculated by the embedded or coated microstrip formulas [33, 34]. Though there is no closed-form solution for the effective dielectric constant, it can be estimated as

$$\varepsilon_{eff} = \frac{\varepsilon_r + \varepsilon_0}{2} + \frac{\varepsilon_r - \varepsilon_0}{2\sqrt{\left(1 + 12\dfrac{H}{W}\right)}}$$

where H is the height of the dielectric, W is the width of the stripline (antenna wing), and ε_r is the dielectric constant of the passivation. The latter value is estimated by the average value of the different layers that build up the passivation. In our cases, the passivation height between the antenna and the lower metal shields is approximately $H = 7$ μm, the width of the antenna wings is $W = 15$ μm, and $\varepsilon_r = 3.8$. The target frequency has been 360 GHz. The estimated antenna L_R size, using the half-wavelength dipole equation, becomes:

$$L_R = \frac{L_0}{\sqrt{\varepsilon_{eff}}} = \frac{\lambda_0}{2\sqrt{\varepsilon_{eff}}} \sim \frac{833 \ \mu m}{2\sqrt{3.3}} = 228.5 \ \mu m$$

Finally, the antenna has been drawn to $L = 230$ μm after EM simulation-based corrections. Figure 5.15 shows the simulated and measured response of this sensor, which show good correlation. The peak amplified responsivity is found to be 185 kV/W @ 365 GHz (1.85 kV/W unamplified responsivity) with open drain configuration and $V_{GS} = 0.2$V (see Figure 5.7). At the detectivity maximum ($V_{GS} = 0.36$V), though the responsivity drops to near half the above values, the NEP reaches 40 pW/$\sqrt{Hz}$.

5.7.3.2 Broadband Antenna

The case of the bow tie antennas is quite different and much more difficult to estimate by paper-and-pencil methods. Beneath the bow tie antennas there is no metal ground shield. Considering the dielectric around the antenna and the substrate with the metallic bottom plate of the packaging, this case can be more precisely modeled as a suspended microstrip structure with superstrate [34]. In order to find the physical dimensions for the resonance frequency, the effective relative permittivity must be derived first, which can be found in the literature as well [33]. In order to make better use of the silicon area, the four-antenna detector pair is connected in series [14].

The difficulties arise from the substrate resonance, as the inevitable Fabry-Pérot interference will amend the ideal behavior. The typical wafer thickness of commercial CMOS technologies varies between 200 and 300 μm. Taking into account the high refractive index of the silicon, this thickness is comparable to the wavelength of

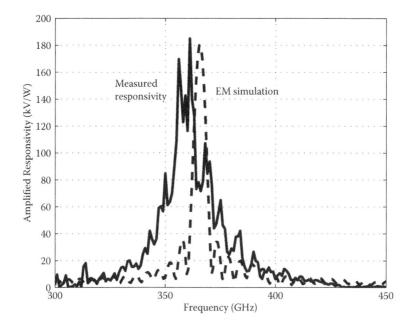

FIGURE 5.15   Simulated and measured responsivity of the H-shaped dipole antenna sensor.

the sub-THz radiation in silicon. During hand calculations, the targeted frequency has been the first significant water absorption peak at 550 GHz. Based on the substrate-air half-sphere model, the resonant size would be $L = 108$ μm. This peak is strongly shifted when the actual dielectric-substrate model and the very close other antennas are involved (Figure 5.16, top section). As a next step, the finite substrate is modeled. The substrate thickness at this technology is $H_{Si} = 270$ μm. This thickness with a metallic bottom plate would give minimum absorption (maximum antenna response) near 220, 380, 540, and 720 GHz and significant attenuation in between. Though the dielectric under the antenna and other metallic structures on the chip alter the simulated peaks, the resulting characteristic presented in the bottom section of Figure 5.7 clearly shows the measured effect of the substrate resonance. As a conclusion, the hand calculations did not conduct useful estimation compared to the successful former case, though detailed EM simulation helped to estimate the real behavior.

The sensor provides 52 kV/W amplified responsivity at 0.47 THz. At $V_{GS} = 0.2V$ the total thermal noise of the four detector is higher than that of the single detector, which results in $NEP \approx 540$ pW/$\sqrt{Hz}$. The peak detectivity needs different settings, namely, $V_{GS} = 0.4V$ with 22 kV/W amplified responsivity and $NEP \approx 120$ pW/$\sqrt{Hz}$.

5.8   APPLICATION EXAMPLES

5.8.1   TRANSMISSION IMAGING

The first example is transmission imaging. The imaging setup (Figure 5.17) is based on reflective elements to allow broadband operation. High-resistivity floating

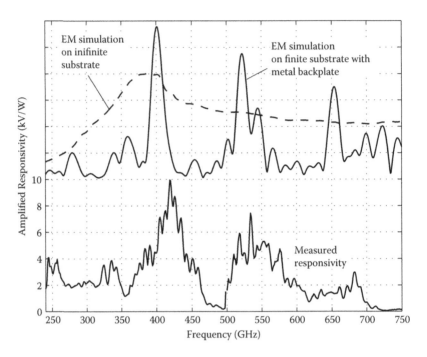

FIGURE 5.16 Simulated and measured responsivity of the four serially connected bow tie antenna sensors. The theoretical smooth and broadband response of the bow tie antenna is changed significantly due to the substrate resonance and the surrounding metal structures as shown in the upper part of the figure.

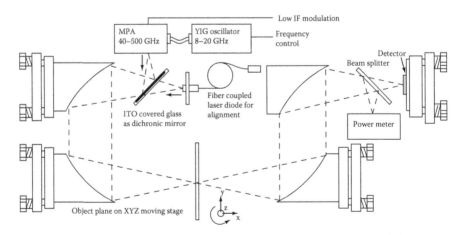

FIGURE 5.17 Transmission image acquisition setup based on a continuous wave (CW) sub-THz source and a quasi-optical arrangement. During the setup process, the high-resistivity silicon beam splitter is removed to aim positioning at the object plane.

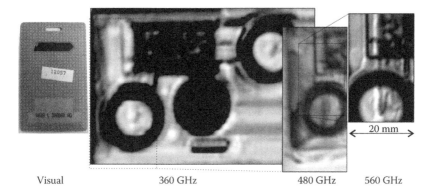

| Visual | 360 GHz | 480 GHz | 560 GHz |

FIGURE 5.18 Transmission raster scan of an active RF entry card at different frequencies captured with different sensors.

zone silicon (HRFZ-Si) beam splitters are used. In order to aid the positioning, an ITO covered glass is placed as a dichroic mirror. During characterization, the absolute power meter is used, while during image acquisition its beam splitter is removed.

In the sample holder an entry card has been fixed, which is a battery-powered transmitter encapsulated into a plastic enclosure. At three wavelengths, the entire portion of the card has been raster scanned. The operation frequencies are selected as peaks of different antennas on the chip: at 360, 480, and 560 GHz the H-shaped, double-H-shaped, and bow tie antenna-coupled sensors were used, respectively. The scanning rasters were 0.5, 0.4, and 0.25 mm for the scans in both horizontal and vertical directions. The total scanned area was 80 × 55 mm at the lowest frequency, and smaller at higher frequencies. The pixel acquisition time varied from 20 to 1 Hz for different wavelengths in order to maintain the image signal-to-noise ratio (SNR), as the corresponding resonant antennas have different sensitivities. The visual and the scans can be seen in Figure 5.18.

5.8.2 COMPLEX WAVEFORM DETECTION

As an application example, a homodyne mixing setup and results are presented. The FET detectors are capable of homodyne or heterodyne mixing according to [12, 13]. These modes help to improve the acquisition SNR. The homodyne detection is based on the self-interference of a coherent wave, which in our case is provided by the electronic multiplier-based source. The source wave is divided into an object and a reference beam, and their interference is captured on the chip surface. The common problem with homodyne imaging is that the interference pattern contains information of the phase difference and the intensity of the object, but this information cannot be separated by a single recording. A usual solution is applying the mechanical delay stage to sweep the reference at least a wavelength to capture phase relations and derive intensity value. This method is inherently slow.

A single-shot quadrature phase-shifting interferometry architecture is presented in [32] that is capable of recording multiple phase-shifted interference values without mechanical translation. The method is based on orthogonally polarized object and reference beams and on linear and circular polarization-sensitive antennas in space-division multiplexing. The method handles the FETs as a square law detector, the response of which depends on the drain-source and gate-source RF voltage. In mixing, the FET's response is approximated as [5]

$$V_R \cong \eta \frac{-\left(U_{DS}^{THz}\right)^2 - \left(U_{DS}^{THz}\right)^2 + 2\left(U_{DS}^{THz} U_{DS}^{THz}\right)\cos\varphi}{4(U_{GS} - U_{DS})}$$

where ϕ is the phase difference between the two incident coherent waves and the two waves are coupled independently to the gate-source and drain-source terminals. In our circuit, there is only a gate-source connection $\left(U_{GS}^{THz} = 0\right)$ and the antenna provides the wave superposition; hence, the above equation becomes simpler. The simplified equation is equal to the square power detection model:

$$V_R \cong \eta \frac{-\left(U_{GS}^{THz}\right)^2}{4U_{GS}}$$

This gate-source voltage was transformed from the free space waves by the antennas through their antenna resistance. By substituting the incoming object and reference wave powers (P_O, P_R), the antenna resistance (R_A), and the phase difference (ϕ), one can get a similar equation for a linear polarization antenna:

$$V_{R, linear} \cong \eta \frac{-R_A P_O - R_A P_R \pm 2R_A\sqrt{P_O P_R}\cos\varphi}{4U_{GS}} \alpha \pm \cos\varphi$$

where the antenna polarization lies in between the orthogonally polarized beams, as can be seen in Figure 5.19. The circular polarization antenna-coupled detector's

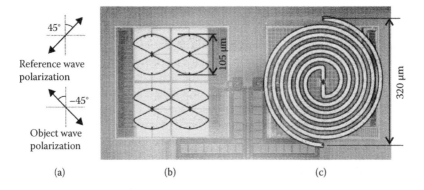

FIGURE 5.19   Microphoto of the bow tie and spiral antenna sensors and the incident reference and object beam polarization.

response will differ from this value, because the circular wings introduce a phase shift in one of the beams coupling to the detector. The sign of the dependence flips using left-handed or right-handed circular polarization:

$$V_{R,circular} \cong \eta \frac{-R_A P_O - R_A P_R \pm 2R_A \sqrt{P_O P_R} \sin \varphi}{4U_{GS}} \alpha \pm \sin \varphi$$

Hence, one can measure two beams independently (i.e., the object and reference beams) and acquire quadrature interferences of 0, $\pi/2$, π, and $3\pi/2$ radians by selecting, aligning, and rotating the linear and changing the handness of circular polarized antenna-coupled detectors.

In the experiment two antennas are selected, a bow tie and an Archimedean spiral, as linear and circular polarized types, respectively. In the optical setup, the reference beam is separated and wire grid polarizers are included as well. The setup is shown in Figure 5.20.

In the presented example an acrylic glass (PMMS) lens has been raster scanned and at each point at 360 GHz. The complex amplitude is recorded as in-phase and quadrature phase interference patterns. The reconstruction of the intensities and phase is straightforward from these data. The exact dimensions of the lens are diameter of 38 mm, focal length of 65 mm @ 360 GHz, and physical thicknesses at the perimeter and center of 2.3 ± 0.05 and 6.5 ± 0.05 mm, respectively. The refractive index of the lens was $n = 1.57 \pm 0.05$ @ 360 GHz. The lens has been placed in the translation stage and the interferograms are recorded in a 30-by-30 mm area with 0.5 mm step size. After reconstructing the intensity and phase, the phase image is unwrapped using a standard method [39]. The measurement steps are shown in Figure 5.21.

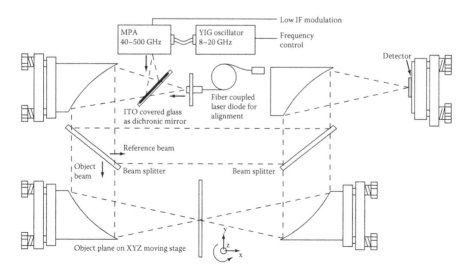

FIGURE 5.20   Homodyne interference image acquisition setup.

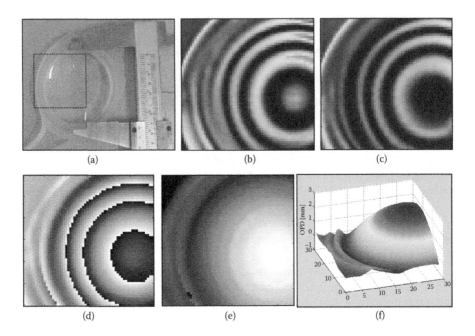

FIGURE 5.21 (a) Photograph of the lens with the raster scanned area of 30 by 30 mm. (b) In-phase interferogram captured by the linear antenna detector. (c) Quadrature interferogram captured by the circular polarization antenna detector. (d) Reconstructed phase image. (e) Unwrapped phase image. (f) Pseudoview of the estimated optical path delay.

5.9 CONCLUSIONS

It has been shown how a versatile imaging system has been integrated around the very same detector. The resulting system autonomously tracks an externally modulated signal and finally provides a compact digital stream containing the measured intensity values of the selected sensors. The practical examples showed that instead of placing an array of identical sensors within the diffraction-limited spot, the integration of different antenna structures yields more interesting and useful applications: broadband response from 0.25–0.7 THz with different polarization sensitivities.

ACKNOWLEDGMENTS

This research project would not have been possible without the support of many people. The author expresses his gratitude to Domonkos Gergely for help in measurements and characterization, Zsolt Benedek for strictly following testability rules, Csaba Füzy and Gergely Károlyi for help in antenna design and coupling, and Tibor Berceli and Ákos Zarándy for supporting the project.

REFERENCES

1. M. Tonouchi, Cutting-edge terahertz technology, *Nat. Photonics*, 1(2), 97, 2007.
2. M.I. Dyakonov and M.S. Shur, Plasma wave electronics: novel terahertz devices using two dimensional electron fluid, *IEEE Trans. Electron Dev.*, 43, 1640, 1996.
3. W. Knap, F. Teppe, Y. Meziani, N. Dyakonova, J. Lusakowski, F. Boeuf, T. Skotnicki, D. Maude, S. Rumyantsev, and M.S. Shur, Plasma wave detection of sub-terahertz and terahertz radiation by silicon field-effect transistors, *Appl. Phys. Lett.*, 85(4), 675, 2004.
4. M. Dyakonov and M. Shur, Detection, mixing, and frequency multiplication of terahertz radiation by two-dimensional electronic fluid, *IEEE Trans. Electron. Dev.*, 43(3), 380, 1996.
5. F. Teppe, Y.M. Meziani, N. Dyakonova, J. Lusakowski, F. Boeuf, T. Skotnick, D. Maude, S. Rumayantsev, M.S. Shur, and W. Knap, Terahertz detectors based on plasma oscillations in nanometric silicon field effect transistors, *Phys. Stat. Sol.*, 2(4), 1413–1417, 2005.
6. R. Tauk, F. Teppe, S. Boubanga, D. Coquillat, W. Knap, Y. Meziani, C. Gallon, F. Boeuf, T. Skotnicki, C. Fenouillet-Beranger, D.K. Maude, S. Rumyantsev, and M.S. Shur, Plasma wave detection of terahertz radiation by silicon field effect transistors: responsivity and noise equivalent power, *Appl. Phys. Lett.*, 89(25), 253511, 2006.
7. J. Lu and M. Shur, Terahertz detection by high electron mobility transistor: effect of drain current, in *Twelfth International Symposium on Space Terahertz Technology*, 2001, vol. 1, p. 103.
8. A. Lisauskas, U. Pfeiffer, E. Öjefors, P.H. Bolìvar, D. Glaab, and H.G. Roskos, Rational design of high-responsivity detectors of terahertz radiation based on distributed self-mixing in silicon field-effect transistors, *J. Appl. Phys.*, 105(11), 114511, 2009.
9. W. Knap, M. Dyakonov, D. Coquillat, F. Teppe, N. Dyakonova, J. Lusakowski, K. Karpierz, M. Sakowicz, G. Valusis, D. Seliuta, I. Kasalynas, A. El Fatimy, Y.M. Meziani, and T. Otsuji, Field effect transistors for terahertz detection: physics and first imaging applications, *J. Infrared Millim. Terahz. Waves*, 30, 1319, 2009.
10. F. Schuster, D. Coquillat, H. Videlier, M. Sakowicz, F. Teppe, L. Dussopt, B. Giffard, T. Skotnicki, and W. Knap, Broadband terahertz imaging with highly sensitive silicon CMOS detectors, *Opt. Express*, 19(8), 7827, 2011.
11. M. Sakowicz, M.B. Lifshits, O.A. Klimenko, F. Schuster, D. Coquillat, F. Teppe, and W. Knap, Terahertz responsivity of field effect transistors versus their static channel conductivity and loading effects, *J. Appl. Phys.*, 110(5), 054512, 2011.
12. D. Glaab, S. Boppel, A. Lisauskas, U. Pfeiffer, E. Ojefors, and H.G. Roskos, Terahertz heterodyne detection with silicon field-effect transistors, *Appl. Phys. Lett.*, 96(4), 042106, 2010.
13. T. Loffler, T. May, C. Am Weg, A. Alcin, B. Hils, and H.G. Roskos, Continuous-wave terahertz imaging with a hybrid system, *Appl. Phys. Lett.*, 90(9), 091111, 2007.
14. T. Elkhatib, V. Kachorovskii, W. Stillman, D. Veksler, K. Salama, X. Zhang, and M. Shur, Enhanced plasma wave detection of terahertz radiation using multiple high electron-mobility transistors connected in series, *IEEE Trans. Microwave Theory Tech.*, 58, 331, 2010.
15. D. Perenzoni, M. Perenzoni, L. Gonzo, A.D. Capobianco, and F. Sacchetto, Analysis and design of a CMOS-based terahertz sensor and readout, *Proc. SPIE*, 7726, 772618, 2010.
16. E. Öjefors, U. Pfeiffer, A. Lisauskas, and H. Roskos, A 0.65 THz focal-plane array in a quarter-micron CMOS process technology, *IEEE J. Solid-State Circuits*, 44(7), 1968, 2009.

17. F. Schuster, D. Coquillat, H. Videlier, M. Sakowicz, F. Teppe, L. Dussopt, and W. Knap, Broadband terahertz imaging with highly sensitive silicon CMOS detectors, *Opt. Express*, 19(8), 7827, 2011.
18. E. Öjefors, N. Baktash, Y. Zhao, and U. Pfeiffer, Terahertz imaging detectors in a 65-nm CMOS SOI technology, in *Proceedings of 36th European Solid-State Circuits Conference*, 2010, p. 486.
19. G. Trichopoulos, L. Mosbacker, D. Burdette, and K. Sertel, A broadband focal plane array camera for real-time THz imaging applications, *IEEE Trans. Antennas Propagation*, 99, 1, 2013.
20. T. Reck, J. Siles, C. Jung, J. Gill, C. Lee, G. Chattopadhyay, I. Mehdi, and K. Cooper, Array technology for terahertz imaging, *Proc. SPIE*, 8362, 836202-1, 2012.
21. P. Földesy, D. Gergelyi, Cs. Füzy, and G. Károlyi, Test and configuration architecture of a sub-THz CMOS detector array, IEEE in *15th International Symposium on Design and Diagnostics of Electronic Circuits and Systems (DDECS)*, 2012, p. 101.
22. P. Földesy and A. Zarándy, Integrated CMOS sub-THz imager array, in *13th International Workshop on Cellular Nanoscale Networks and Their Applications*, 2012, p. 1.
23. J. Van Rudd and D. Mittleman, Influence of substrate-lens design in terahertz time domain spectroscopy, *JOSA B*, 19, 319, 2002.
24. H. Sherry, R. Al Hadi, J. Grzyb, E. Öjefors, A. Cathelin, A. Kaiser, and U. Pfeiffer, Lens-integrated THz imaging arrays in 65nm CMOS technologies, in *IEEE Radio Frequency Integrated Circuits Symposium (RFIC)*, 2011, p. 1.
25. F. Friederich, W. von Spiegel, M. Bauer, F. Meng, M. Thomson, S. Boppel, A. Lisauskas, B. Hils, V. Krozer, A. Keil, T. Loffler, R. Henneberger, A. Huhn, G. Spickermann, P. Bolivar, and H. Roskos, THz active imaging systems with real-time capabilities, *IEEE Trans. Terahertz Sci. Technol.*, 1, 183, 2011.
26. G. Trichopoulos, G. Mumcu, K. Sertel, H. Mosbacker, and P. Smith, A novel approach for improving off-axis pixel performance of terahertz focal plane arrays, *IEEE Trans. Microwave Theory Tech.*, 58, 2014, 2010.
27. R.R. Harrison and C. Charles, A low-power low-noise CMOS amplifier for neural recording applications, *IEEE J. Solid-State Circuits*, 38(6), 958, 2003.
28. Z.Y. Chong and W.M.C. Sansen, *Low-noise wide-band amplifiers in bipolar and CMOS technologies*, Springer International Series in Engineering and Computer Science, Springer, Berlin, 1991.
29. J. Kim and M.A. Horowitz, Adaptive supply serial links with sub-1-V operation and per-pin clock recovery, *IEEE J. Solid-State Circuits*, 37(11), 1403, 2002.
30. P. Földesy, Array of serially connected silicon CMOS sub-terahertz detectors per pixel architecture, in *3rd EOS Topical Meeting on Terahertz Science and Technology (TST 2012)*, Prague, Czech Republic, June 2012.
31. P. Földesy, Z. Fekete, T. Pardy, and D. Gergelyi, Terahertz spatial light modulator with digital microfluidic array, in *Procedia Engineering: 26th Eurosensors Conference*, 2012, vol. 47, p. 965.
32. P. Foldesy, Terahertz single-shot quadrature phase-shifting interferometry, *Optics Lett.*, 37(19), 4044, 2012.
33. P. Sullivan and D. Schaubert, Analysis of an aperture coupled microstrip antenna, *IEEE Trans. Antennas Propagation*, 34(8), 977, 1986.
34. R. Garg, *Microstrip antenna design handbook*, Artech House Publishers, Norwood, ME, 2001.
35. R. Johnson and H. Jasik, *Antenna engineering handbook*, McGraw-Hill Book Company, New York, 1984.
36. P. Maraghechi and A. Elezzabi, Experimental confirmation of design techniques for effective bow-tie antenna lengths at THz frequencies, *J. Infrared Millim. Terahz. Waves*, 1, 2011.

37. L.T. Wang, C.E. Stroud, and N.A. Touba, *System-on-chip test architectures: nanometer design for testability*, Morgan Kaufmann, Burlington, MA, 2008.
38. Y. Zorian, E.J. Marinissen, and S. Dey, Testing embedded-core based system chips, in *IEEE International Test Conference*, 1998, p. 130.
39. J.M. Huntley, Noise-immune phase unwrapping algorithm, *Appl. Optics*, 28(16), 3268, 1989.

6 40/100 GbE Physical Layer Connectivity for Servers and Data Centers

Yongmao Frank Chang[*]

CONTENTS

---

[*] Currently affiliated with Inphi Corporation, Westlake Village, California.

6.1   INTRODUCTION

The rapid growth of Internet, cloud computing, and mobile networking applications has resulted in data center network bandwidth requirements that outpace Moore's law [1]. This exponential growth in core, metro, and access bandwidth will likely continue for many years to come. Figure 6.1 shows the data center architecture based on a proven layered network design. A typical three-layer (or tier) data center network comprises an access layer, an aggregation layer, and a core layer. Currently a large number of 10 Gbps networks have already been laid by service providers. Mega data center operators are bundling multiple 10 GbE links in parallel via link aggregation (LAG) [2]. However, this approach not only is complex to configure (with too many cables to manage), but also can cause load imbalance across multiple LAG links, making it harder to scale. Pooling multiple server data streams into a LAG group can improve utilization, but it can also destroy the natural parallelism of the data stream and introduce performance latency. Figure 6.2 shows the bandwidth trends for data center networking and server I/O [3]. Networking interconnect bandwidth must scale

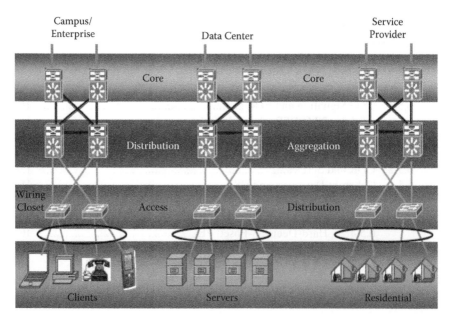

FIGURE 6.1   Data center architecture overview with a proven layered approach. (Courtesy of Cisco.)

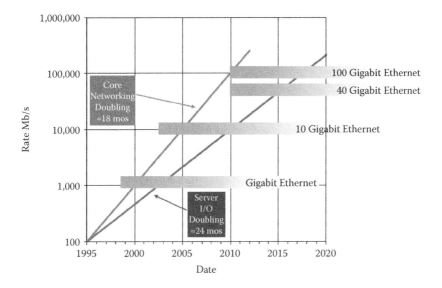

FIGURE 6.2   Bandwidth trends for data center networking and server I/O. (From IEEE 802.3 HSSG Tutorial 1107, November 2007, http://www.ieee802.org/3/hssg/public/nov07/HSSG Tutorial 1107.zip.)

from 10 Gb/s to 40 Gb/s and then to 100 Gb/s, while the server interconnect at the bottom of the data center hierarchy is moving up to 10 Gb/s at a much slower pace, to 40 Gb/s for the next few years.

This predicted ever-increasing demand for both enterprise and carrier-grade 40/100 G Ethernet systems in the near future has led to significant research and development, as well as standardization efforts [4–7] regarding various physical layer interface options. Depending on targeted deployment scenarios, one approach uses parallel implementation, by splitting 100 Gb/s traffic among multiple parallel lanes (i.e., in single-mode fiber cables or multimode fiber ribbons) or among multiple multiplexed wavelengths. Another approach implements 100 Gb/s in a serial fashion (i.e., on a single optical wavelength and a single fiber), with wavelength-division multiplexing (WDM) as an option to carry multiple 100 Gbps data streams. While parallel options are expected to dominate short-reach applications in enterprise, access, and interconnect space, the serial transport is viewed as the most cost-effective choice for 100 G transport in provided networks for the sake of spectral efficiency and reach objectives. The adaption of advanced modulation schemes such as dual-polarization quadrature phase shift keying (DP-QPSK) signaling with coherent receivers for regional and long-haul applications is driving significant investment into this area.

Fiber optic technologies play critical roles in data center and server connectivity. Optical transceivers and transponders (transmitters and receivers) are perceived as significant components of the cost in building networks. A transponder and transceiver are both functionally similar devices that convert a full-duplex electrical signal in a full-duplex optical signal. The difference between the two is that transceivers interface electrically with the host system using a serial interface, whereas transponders use a parallel interface to do so. So transponders provide easier-to-handle lower-rate parallel signals, but are normally bulkier and consume more power than transceivers.

For data centers and servers, there is tremendous pressure to develop cost-effective, high-performance, and compact transceivers that operate under real-world and practical constraints. This chapter describes the building blocks, their associated integrated circuit (IC) technologies, especially in the physical layer, various multi-source agreement (MSA) formats, system architectures, and related issues of such transmitters and receivers, or simply transceivers for 100 G technologies together with relevant 40 G technologies (and beyond).

It's worthwhile to mention that as building blocks of critical enabling technology, the demand for various physical layer IC devices is becoming urgent [8, 9], and their stable development will help accelerate the transition to 40/100 G deployment.

6.1.1 Historical Background

Ethernet was invented by Metcalfe and others at Xerox's Palo Alto Research Center in the mid-1970s with an experimental rate of 3 Mbps [10]. It was commercially introduced in 1980 and standardized in 1985 as IEEE 802.3. Ethernet initially competed with two largely proprietary systems: Token Ring and Token Bus. Because Ethernet was able to adapt to market realities and shift to inexpensive and ubiquitous twisted pair wiring, these proprietary protocols soon found themselves competing in a market inundated by Ethernet products, and by the end of the 1980s, Ethernet was clearly the dominant network technology. Since then, Ethernet technology has evolved to meet new bandwidth and market requirements. Data rates were periodically increased from the original 10 Mb/s to 10 Gb/s and now to 40/100 Gb/s, with 1 terabit under discussion nowadays.

The Institute of Electrical and Electronics Engineers (IEEE) first started project 802 to standardize local area networks (LANs) in February 1980. The so-called Blue Book carrier sense multiple access with collision detection (CSMA/CD) specification was first submitted as a candidate for the LAN specification [11] by the "DIX group," formed by representatives from DEC, Intel, and Xerox [12]. In addition to CSMA/CD, Token Ring (supported by IBM) and Token Bus (supported by General Motors) were also considered candidates for a LAN standard. Competing proposals and broad interest in the initiative led to strong disagreement over which technology to standardize. In December 1980, the group was split into three subgroups, and standardization proceeded separately for each proposal with significant delays in the standards process.

Later a broader support for Ethernet beyond IEEE was quickly achieved by the establishment of a competing task group, Local Networks, within the European standards body ECMA TC24. As early as March 1982, ECMA TC24, with its corporate members, reached an agreement, in cooperation with IEEE 802, on a standard for CSMA/CD based on the IEEE 802 draft. Because the DIX proposal was the most technically complete and because of the speedy action taken by ECMA, which decisively contributed to the conciliation of opinions within IEEE, the IEEE 802.3 CSMA/CD standard was approved in December 1982. IEEE published the 802.3 standard as a draft in 1983, and ratified it as a standard in 1985.

Ethernet soon showed many advantages that included cost efficiencies, unprecedented scalability and flexibility, protocol neutrality, ease of use, and reliability over other legacy protocols. The ubiquity of Ethernet has enjoyed dominance in local area networks (LANs) for the private and enterprise sectors, and starting in the 1990s, it migrated beyond its initial LAN domain into much larger areas, such as metropolitan area networks (MANs) and wide area networks (WANs).

The IEEE has defined a distinct set of physical layer data rates for Ethernet with a set of interface options (either electrical or optical). An Ethernet physical layer moves signals such as the 10 GbE WAN signal over the optical transport network (OTN) or 1 GbE signal over the synchronous optical network (SONET) using transparent Generic Framing Procedure (GFP) mapping, transparent over a public transport network. Ethernet physical layer connections are point to point only, and always operate at the standardized data rates.

As shown in Figure 6.3, the industry activity for 100 G deployment includes both Ethernet and transport. It is worthwhile to note that the Ethernet and transport standard are converging at a uniform speed of 100 Gb/s [8], a process that started at ~10 Gb/s.

Data centers and commercial network operators have cited the need for 100 GbE connectivity and transport to support ever-increasing Internet traffic growth over the next decades. The industry is responding to these demands for more bandwidth as evidenced by the collective standardization efforts within the IEEE 802.3 Working Group, ITU-T Study Group 15, and OIF Physical Layer Group to define 40/100 GbE, OTU4, and 100 G transport specifications as illustrated in Figure 6.4. With these standards well established, systems, subsystems, modules, and components vendors have formed efficient industrial ecosystems toward cost-effectively deploying a new generation of optical transceivers for 100 GbE client interconnection and line system transmissions.

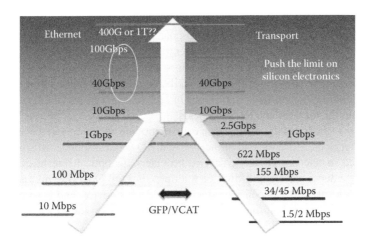

FIGURE 6.3   Convergence of Ethernet and transport that started at ~10 Gb/s rates.

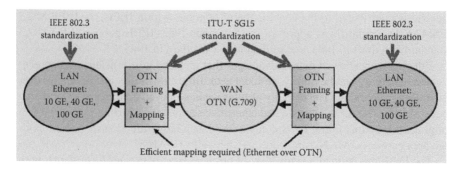

FIGURE 6.4   Ethernet drives the demand for next-generation transport systems.

6.1.2 TRANSCEIVER EVOLUTION

Transceivers act as the all-in-one objects that transmit, receive, and convey data information, similar to transmitters and receivers found in radio systems. Using an optical transceiver, networks save more space and avoid the need to have a transmitter and receiver inside a network. The newer transceivers are normally capable of transmitting further and faster, and at the same time are smaller, more compact modules than older models.

The earliest transceivers were created for Gigabit Ethernet networks and were preferred for their hot-swappable abilities similar to the newer SFP+ (small form factor pluggable) modules that are well know today. Gigabit interface converters (GBICs) allowed networks the ability to transmit data across copper or fiber optic channels, creating a more versatile device than transmitters and receivers. Of course, GBIC modules were not without their imperfections, and many had size and compatibility issues that limited their ability to transmit data across particular distances and at certain wavelengths.

As we know well as of today, optical transceivers have gone through a series of radical evolutions for the last two decades. Optical modules are highly integrated small package developments of optical components and physical layer integrated circuits. Highly integrated devices help the user to deal with high-speed analog signals, shorten the R&D and production cycle, reduce the vitality of types of procurement, reduce production costs, and therefore facilitate ramp-up of the volume deployment. Currently over hundreds of millions of optical transceivers have already been shipped into the telecom and datacom fields.

Figure 6.5 illustrates the comparison of 10 G module evolution and the associated module interfaces. Since the end of the 1990s, a 300-pin transponder MSA has been widely used in telecom space by service providers for optical interfaces of high-performance dense wavelength-division multiplexing (DWDM) systems and routers. The 10 G 300-pin MSA is considered a cold pluggable device, and is widely available only for 1550 nm long/intermediate-reach and 1310 nm intraoffice/short-reach applications. It contains electrical (de)multiplexing functions to aid system designers in managing the signal integrity at the system circuit level, i.e., a 16-bit multiplexing function with lower-speed electrical interface at 622 Mb/s. A standard around

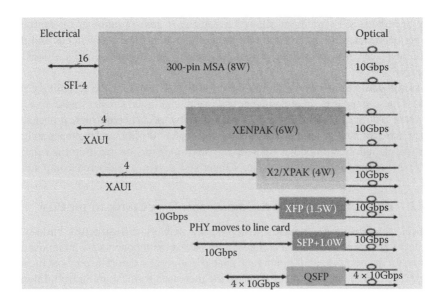

FIGURE 6.5   10 G module evolution and its electrical and optical interfaces. (Courtesy of LightCounting.)

this electrical interface and signal management was developed by OIF, commonly referred to as the SFI4-1 specifications [9]. With response to the optical interface, the 30-pin modules are capable of conveying data across short and long distances due to their configuration settings located inside the devices. A 300-pin MSA contains a full description of the mechanical aspect of the module package and connectors. Initially a maximum footprint was typically 5×5 inch^2. Later, a small form factor version was reduced to 3.5×4.5 inch^2 and even compact as 2.2×3 inch^2.

In the datacom space, with increased support across longer fiber distances and for multiple wavelengths, XENPAK transceivers became the new standard to upgrade over the GBIC transceiver. Equivalent to the 300-pin transponder, XENPAK transceivers were capable of conveying data across short and long distances. XENPAK transceivers leverage the same "hot pluggable" feature that was available with the GBIC, as well as the SFP modules that covered lower data rates below 10 Gb/s. XENPAK has a size of approximately 2×5 inch^2 with an industry standard 70-pin electrical connector, of which the input and output data signals are transmitted per a new electrical interface specification, called XAUI, as defined by IEEE 802.3ae. With the 10 GbE standard taking hold, XENPAK transceivers were soon migrated into the newer X2 and XPAK modules. The smaller, more flexible X2 and XPAK standards allowed for even more support for the different Ethernet standards and were capable of transmitting data across longer distances. The X2/XPAK transceivers are still shipping well in volume quantity today, mostly for router and switch interfaces in enterprise data centers.

When the serial newer XFP and SFP+ modules came into existence for 10 Gb/s, the competing forms of X2 and XPAK could not continue to control the market as they once had. SFP+ modules allowed greater *port density* (number of transceivers per centimeter along the edge of a motherboard) in Ethernet switches and network interface cards, providing various wavelength and distance configurations for Ethernet.

As standards transform from 10 Gb/s to 40/100 G, so does the technology that utilizes these standards, creating higher bandwidth and faster, smaller transceivers for networks to utilize in sending information across networks better and more proficiently. A new series of next-generation optical transceivers for 100 GbE and its transport have been developed. Such modules allow the system designers to speed up their design cycle, saving time and resources vs. implementing a design with discrete components.

6.1.3   100 G Transceiver Applications from Data Center to the Core

Optical transceivers are needed for high-speed network infrastructure build-outs, for both carriers and data centers. Optical transceiver components are an innovation engine for the network, and support and enable low-cost transport throughout the network. The market driving forces relate to the increased traffic coming from Internet video, cloud-based services, and mobile connectivity. All those associated devices are networked and drive significant traffic to the broadband network, stimulating the demand for optical transceivers. The global optical transceiver markets are poised to achieve significant growth as the data in networks expand exponentially. It is estimated the market for optical transceivers will grow to ~$7 billion by 2020, driven by the availability of 100 Gbps devices and the vast increases in Internet data traffic.

The optical transceiver signal market is intensely competitive and relatively cost-effective. There is increasing demand for optical transceivers as communications markets grow in response to more use of smart phones and more Internet transmission of data. The market for network infrastructure equipment and for communications semiconductors offers attractive long-term growth.

Figure 6.6 shows one example of a typical optical network from data centers to the core. Optical networks provide transport, multiplexing, switching, management,

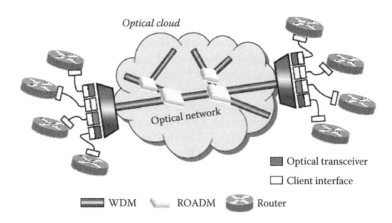

FIGURE 6.6   A high-level view of typical 40/100 G optical WDM networks.

supervision, and survivability of communication channels. Routers and switches with 10 G and 40 G interfaces emerged to hand off to DWDM transmission terminals, driving the need for 40 G wavelengths. Router handoffs will ultimately push the need for 100 G as access streams into routers become higher data rates (10 G and higher) that force the transport backbone link to go to an even higher rate at 100 Gbps and beyond.

Changes in network architecture are another driver for higher-bit-rate networks. In the past, providers had separate networks for each access technology or service. Today, most providers are attempting to put (aggregate) these services and technologies onto one Internet Protocol (IP) backbone, which puts pressure on the core backbone to handle more traffic than ever. For example, when AT&T merged multiple networks (SBC and BellSouth), it led to integrated networks overflowing onto one large backbone that required an immediate upgrade to 40 G and is now 100 G capable.

Consolidating traffic onto fewer wavelengths and the associated economics were other factors for the transition in the drive toward faster optical channel rates. Larger-bandwidth wavelengths have always promised better operation efficiencies merely because there are fewer wavelengths to manage and a smaller number of parts in the network that can fail.

The years 2011 and 2012 witnessed the significant ramp-up in commercial deployment of 100 G technologies, and the research community has started to look into technologies beyond 100 G [13, 14]. The 100 G deployment can be clarified into two major areas: the client applications for LAN in the enterprise and the line application for the WAN in backbone networks.

In 100 GbE client applications, IEEE P802.3 has carried out the standardization of 100 GbE by incorporating 40 GbE to address the near-term needs of the server and storage community for low-cost Ethernet interfaces beyond 10 G. The proposed 100 G module interfaces span from copper media of 7 m, to multimode fiber media of 100 m, to single-mode fiber media up to 40 km, and all use parallel bit streams. The copper applications specify 10 lanes, each operating at 10 Gb/s transmitted over 10 copper pairs for each direction, assembled into a single cable. The multimode fiber applications specify 10 lanes, each operating at 10 Gb/s transmitted over 10 individual fibers for each direction, typically assembled into a single-ribbon cable. All these options leverage the mature and cost-effective 10 Gbps technology. The single-mode fiber applications specify four lanes at 25 Gb/s transmitted over a single duplex fiber at four different wavelengths around 1300 nm. Related to these choices of parallel transmission is the need for the novel 10:4 noninteger multiplexing or "gearbox" and subchannel deskew functionality.

Over the line side, OIF has developed a framework that identifies the target system objectives, functional architectures, and technology building blocks about optical transceivers for 100 G DWDM long-haul transport. The implementation of 100 G optical transceivers includes the modulation format, coherent detection, forward error correction (FEC) considerations, electromechanicals, and management interface. OIF adopted dual-polarization quadrature phase shift keying (DP-QPSK) with coherent detection for specifying the optical transmitter and receiver designs. One key benefit is DP-QPSK reduces the required symbol rate by a factor of four, thus enabling lower-cost optoelectronic technologies to be utilized for 100 G transmission, and in the meantime tolerating various optical propagation impairments, such as chromatic and polarization-mode dispersion and optical signal-to-noise ratio (OSNR).

It's worthwhile to mention that the choice of those module interfaces results from many thorough debates and trade-offs by the industrial community on market requirements for port density, latency, spectral latency, etc., while still maintaining superb performance for each served market.

6.2 MARKET DRIVERS AND REQUIREMENTS

Optical networks demand ever-increasing data rates for enterprise and carrier applications, with the primary driver coming from the end users' ceaseless desire for high bandwidth, in particular, because nowadays everything is going mobile. This evolution is driven by mobile smart phones and tablets that provide universal connectivity. So there is expected to be tremendous investment in wireless cell tower base stations as the quantity of network traffic grows exponentially. It comes as no surprise that such traffic growth is the underlying driver for 100 Gb/s backbone networks. This was the case with the transition from 2.5 G to 10 G, and 10 G to 40 G, and now 100 G optical channel rates.

Carriers worldwide are responding to the challenges brought by the massive increases in wireless data traffic. The advent of big data and exponential growth of data managed by the enterprise data centers are significant market factors. Now in the optical communication network, such as from LAN to MAN to WAN, the optical and electronic devices of optical transceivers are in high demand, to enable more and more optical transceiver module types with increasingly high requirements and complexity. The optical transceiver development direction follows a path toward miniaturization and low-cost development, thus making the optical network configuration more complete and reasonable. With the optical transceiver module's sharp increases in diversity, the industry needs to continue to develop related connectivity technologies that satisfy the application requirements at an amazing speed.

6.2.1 MAXIMIZING PORT DENSITY

One of the major drivers for data centers is associated with *data center network convergence* for server virtualization and interface I/O consolidation, as shown in Figure 6.7. In massive data centers such as Google or Facebook, there are always tens of thousands of servers in the facility, and the growing huge data demand is exhausting the capacities of many data centers, making them easily run out of space, power, or cooling. Server virtualization is one efficient way to tackle this bottleneck by consolidating servers and improving utilization. In server virtualization technologies, one single physical server previously running only one application is now hosting multiple or tens of virtual machines for different applications of multiprotocols. These effectively increase the I/O bandwidth required by a factor of 10 or more, necessary to support multiple hosted applications. As virtual machine density continues to increase on the server, $N \times 10$ GbE connectivity is common and 100 GbE becomes much more attractive.

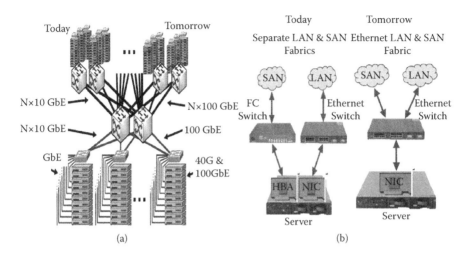

FIGURE 6.7   Data center topology showing the impact of (a) server virtualization and (b) I/O consolidation for data center network convergence.

Server virtualization is changing the requirements for data center networking, and more importantly, for storage networks with the wave of resource consolidation. The cloud computing era introduced a decentralization of compute resources, while the traditional storage area network (SAN) runs on dedicated networks. As this network continues to grow, the costs and resources to deploy and manage SAN begin to compete for the costs from LAN. And LANs are ubiquitous and never going away. So, as Ethernet continues to increase in capability to support Fibre Channel traffic with FCoE (Fibre Channel over Ethernet), the opportunity to consolidate both the LAN and the traditional SAN becomes a must. This transition has started from 10 GbE, and is being realized in 40/100 GbE.

100 GbE offers the required bandwidth for converged traffic while also introducing various significant economic benefits. Data center networking is extremely cost-sensitive, while cabling represents a significant portion of a typical data center budget. With increased performance (throughput) associated with 100 GbE, fewer adapters, fewer cables, and fewer switch ports are required to support the same data traffic of previous 10 GbE products. Price reduction for 40/100 GbE is now at the point where the cost per gigabit of bandwidth is going to be much less for 100 GbE vs. 10 GbE. Per port costs for 40/100 GbE switches are dropping rapidly as demand for 40/100 GbE is now driving volume.

Power consumption is another prominent consideration for data center networking. There are many green initiatives to provide some significant economic advantages to deploying new, greener technologies. Server virtualization helps reduce the number of physical servers needed for the increased I/O requirements. Consolidating onto 100 GbE from 10 GbE reduces the number of adapters, cables, and switch ports required to support the same I/O requirements. Reduction in equipment translates into less power and cooling requirements, in addition to the reduction in equipment costs.

To meet the various requirements for data center upgrades with a transition of server connectivity from 10 GbE to 40/100 GbE, equipment vendors place greater emphasis on the port density of LAN switches and routers as a differentiating factor among competing products. The port density is defined as the number of ports to support by a network switch, router, or hub. The more ports (the greater the port density), the more devices or lines, or traffic throughput that can be supported by the switch. As one example, the current high-density top of rack (TOR) access switch has 48 × 10 GbE SFP+ ports of 480 Gb/s throughput for server connectivity [15].

Low-cost and high-speed pluggable optical transceivers enable increased port density in Ethernet and storage equipment. A series of hot pluggable optical transceivers capable of operating over many different physical mediums and at different distances has been developed for intra-data center connectivity. "Hot pluggable" means the module can be hot swappable while the device is operating for ease of removal and replacement. Transceivers are not specified by the IEEE 802.3 standard but by multisource agreements (MSAs) among switch vendors and chip manufacturers. Those 10 GbE modules as shown in Figure 6.8 cover various transceiver types from 300 pin, XENPAK, and X2/XPAK, to XFP and SFP+. Figure 6.9(a) shows the historical evolution of 10 GbE duplex fiber form factors, enabled by increases in integration and electrical I/O rate from parallel to serial. While fewer 10 G form factors might have been possible, this would have resulted in delaying front-panel density increases by not taking advantage of incremental technology advances [16].

The MSA is an ongoing effort, driven by the need to meet demands for higher port densities, lower power consumption, smaller form factors, higher performance, and lower price points. Figure 6.9(b) shows the evolution of 100 G duplex fiber form factors defined by the CFP (C* form factor pluggable) MSA [17]. Increases in front-panel density result from decreases in module size and power. This is enabled by increased integration of optics and ICs, and by an increase in electrical I/O rate from 10 G to 25 G, reducing the I/O width from 10 lanes to 4.

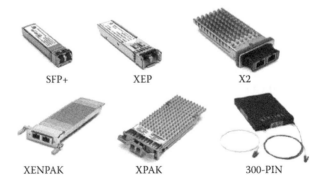

| SFP+ | XEP | X2 |

| XENPAK | XPAK | 300-PIN |

FIGURE 6.8   10 GbE optical transceivers (not drawn to scale). (Courtesy of Finisar.)

---

* C stands for the number 100 (centum in Latin).

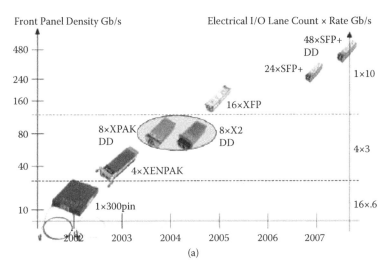

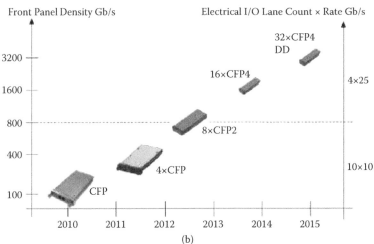

FIGURE 6.9 Optical transceiver module form factor evolution: (a) 10 G modules and (b) 100 G modules.

6.2.2 NETWORK LATENCY

Traditionally, the metric of focus in data center networking has been *bandwidth*. We should say the bandwidth is just one element of what a user perceives as the speed of a network. A lesser known element of network performance—*latency*—also plays an important role. As more and more parts of the Internet are ramping up to have their capacity upgraded, network-induced latency becomes one main problem, which often has noticeable impact on performance.

Network-induced latency, a synonym for delay, measured in one-way delay or round-trip time, is normally defined as how much time it takes for a data packet to travel from one point to another. Several factors contribute to the latency. Besides propagation delays at the speed of light with fiber itself, latency may also involve any transmission delays (properties of the physical medium) in the printed circuit board (PCB) and processing delays (such as passing through proxy servers or making network *hops* on the Internet). Latency increases when any additional equipment layers are added.

For a network connection, the peak of the available bandwidth may be fixed theoretically per the technology adopted, but the actual bandwidth may be affected negatively by high latency. The actual bandwidth can vary over time; excessive latency creates bottlenecks that prevent data from filling into the connection pipe, thus reducing the effective bandwidth of a network. The impact of latency on bandwidth can be temporary (which may last from tens of milliseconds to a few seconds) or persist (in a constant manner), depending on the type of delays.

As new services evolve in data centers, low latency becomes an increasingly important factor in a growing portion of applications and data center projects. Today's most demanding applications, such as transporting video or executing transactions at high speed and in high volume for the financial industry, interconnecting high-performance computing (HPC) clusters via InfiniBand, or enabling synchronous storage services via Fibre Channel and online gaming, are known to be highly sensitive to latency. These emerging applications are driving new requirements for ultra-low-latency data center network designs.

Online gamers are possibly among the Internet users who are most aware of the importance of latency [19]. In real-time multiplayer games over the networks, players don't want to be put into a disadvantageous situation because their actions take longer to reach the game server than those of their components. In the meantime, latency impacts the other users of the Internet as well. At a given bottleneck bandwidth, a connection with lower latency will reach its achievable rate faster than one with higher latency. Latency determines not only how players experience online real-time game play, but also how to design the games to mitigate its effects and meet player expectations.

High-frequency trading is driving new requirements for ultra-low-latency data center network designs. Normally low-latency data are uncompressed, and require more bandwidth [18]. The financial industry is looking for latency numbers to be reduced from milliseconds to microseconds. Most stock exchanges have been busy upgrading their trading platforms. Brokers have tried their best to differentiate their direct market access programs on speed execution, and quite too many ultra-low-latency system vendors have announced and promoted light-speed performance of their gears with low-latency connectivity.

We can discuss how important latency is to high-frequency trading communities in financial markets. As one case study to take advantage of latency, one may have heard of the "latency arbitrage" practice by huge hedge funds in the last several years that have been able to make billions of dollars by picking the pocket of every retail investor. What they do is locate their computers (servers) as close as they possibly can to the electronic exchanges that execute their trades, and they pay exchanges to give them actual stock price information before that raw data get consolidated and sent to most other market players. As those latency arbitrageurs know what the actual prices are, say, 100 to 200 milliseconds ahead of the retail investor, they are able to earn money, making between one and three cents on each share at a time. It was reported by the *Wall Street Journal* that all these pennies add up to $3 billion per year, paying those hedge fund moguls at everyone else's expense.

We would expect that big exchanges like the New York Stock Exchange (NYSE) would try their best to improve the fairness of the playing field for all market participants. Figure 6.10 shows the calculated latencies between neighboring cities for the current NYSE's $500 million, 400,000-square-foot co-location data center in Mahwah, New Jersey [20]. To avoid any latency arbitrage practice, any delay or queuing inserted into the trading path must be eliminated.

Banks want lower-latency service. Because of advanced trading systems and in-house technology, even a 5-millisecond delay is considered an outage. The competitiveness of the market is no longer measured in milliseconds (ms), but in microseconds (μs). The clients must have the latest technology to compete in the ultra-low-latency high-frequency trading market. And it is estimated that 10 milliseconds of latency could potentially result in a 10% drop in revenues for a firm [20, 21]. Therefore, the financial services industry faces tremendous pressure to optimize the transaction life cycle. There is a critical need for the underlying infrastructures to deliver extremely low-latency and very high message throughputs.

For a deep look at the sources of connectivity latency, beyond fiber propagation delays, network equipment can cause delays due to the processing speed for the data traffic, which might incur milliseconds of latency. These physical layer contributions may include clock and data recovery, serialization/deserialization, data encapsulation

Latency data of NYSE Euronet

- Mahwah to Newark or Weehawken

    140 microseconds
    26 line-of-light-miles

- Mahwah to White Plains

    135 microseconds
    25 line-of-light-miles

- Mahwah to Carteret

    216 microseconds
    40 line-of-light miles

FIGURE 6.10 The calculated latencies for NYSE Mahwah data center. (From A. Bach, The Financial Industry's Race to Zero Latency and Terabit Networking, presented at OFC 2010, keynote speech for service provider summit.)

and equalization filters, performance monitoring, protocol conversion, and forward error correction (FEC) algorithm. Best-in-class, purpose-built transceivers and transponders (or muxponders) have emerged that produce latency that can be maintained within less than the high nanoseconds.

6.2.3 SPECTRAL EFFICIENCY

One critical objective of the optical transport systems in long-haul space is to mitigate the networks into the next-generation WDM systems with the ability to transport the growing traffic without replacing the existing fiber cable infrastructures, which would require significant capital investment. A large quantity of fiber cables (either terrestrial or submarine) were installed globally by the year 2000, and such cable infrastructures are scarce and expensive, and thus will continue to be employed during the following decade. The question will be: What's the limit on increasing the transmission capacity over one strand of fiber on the existing fiber plant, or simply, how much capacity can a single-mode fiber support (say, limited to an erbium-doped fiber amplifier (EDFA) C-band)?

In information theory, Shannon's Limit, formulated by Claude Shannon [21], a mathematician who helped build the foundations for the modern communication networks, is a statement that expresses the maximum possible data speed that can be obtained in one data channel. It states that the highest obtainable error-free data speed, expressed in bits per second (b/s), is a function of the bandwidth and the signal-to-noise ratio.

Let C be the maximum obtainable error-free data speed in b/s that a communications channel can handle. Let W be the channel bandwidth in Hz. Let S/N represent the signal-to-noise ratio. Then Shannon's limit is stated as follows:

$$C = W \times \log_2 (1 + S/N)$$

So the information capacity increases linearly with bandwidth, but only sublinearly with S/N. No practical communications system has yet been devised that can operate at close to the theoretical speed limit defined by Shannon's limit.

The fiber link spectral efficiency is defined as C/W, measured in bit/s/Hz. It is the net bit rate (useful information rate excluding error-correcting codes) or maximum throughput divided by the bandwidth in hertz of a communication channel or a data link. In 2000, the widely deployed WDM systems typically transported 80 channels of 10 Gb/s data at 50 G spacing in the EDFA C-band of about 4 THz bandwidth. That is equivalent to 0.2 (b/s)/Hz in spectral efficiency.

It is still sensible from a cost perspective to target the use of the existing fiber infrastructure for high-transmission capacity and spectral efficiency.

There are several ways to increase the capacity of the transport networks: use high data rates, wider amplifier bandwidth, or narrower channel spacing. To increase the data rates over a single wavelength is the most natural first step, as it can be supported by existing link systems. As shown in Figure 6.11, for 50 GHz spaced systems of a total 80 wavelengths to carry 100 GbE traffic per wavelength, the spectral efficiency is 2 (b/s)/Hz, a factor of 10 improvement.

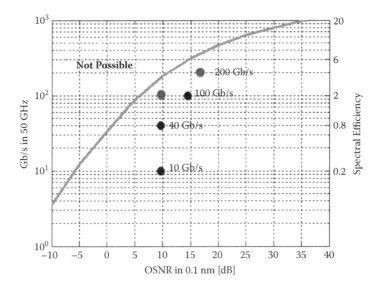

FIGURE 6.11 Shannon's limit and spectral efficiency of WDM systems with regard to OSNR. (Courtesy of Ciena.)

Many carriers have conducted field trials and initial commercial deployments using 100 Gb/s per wavelength [22 and references therein]. Advanced modulation formats and high-end digital signal processing (DSP) are key enablers for 100 G transport. Over several years' debate, the industry has converged on DP-QPSK with coherent detection as the preferred format for 100 Gb/s transport applications. The baud rate of the DP-QPSK 100 Gb/s channel is a quarter of the data rate, so the channel easily fits into a 50 GHz spaced channel plan. DSP- and analog-to-digital converter (ADC)-enabled coherent detection help mitigate fiber impairments such as chromatic dispersion (CD) and polarization-mode dispersion (PMD), in addition to other complex nonlinear transmission impairments.

Such innovative 100 G coherent solutions enable (serial) transport of 100 G data rate capacity over a single wavelength across long-haul distances with higher optical performance than 10 G solutions. The coherent solution can operate over 2500 km without the need for a dispersion compensator. With linear fiber impairments compensated, optical channel performance becomes limited primarily by the OSNR tolerance, as illustrated in Figure 6.11. It's understandable that sophisticated forward error correction (FEC) algorithms are being developed [23].

There consistently exists industrial effort to push toward similar solutions in order that an entire ecosystem of components, subsystems, and system vendors collectively work together to bring products to market quicker and at the best possible cost points. Over time, many complex emerging techniques can be implemented into the transceivers to improve the channel capacity, performance, and cost. However, to reduce the DSP algorithm complexity to achieve a practical implementation, especially for an inroad into the metro space, still represents a challenge, with the aim of significantly reducing size, power consumption, and cost.

As channel data rates have increased, the advanced 100 G transceiver design relies heavily on the integration of optical components and the associated electronics circuit, in other words, the photonics integrated circuits (PICs). The PIC effort is to integrate various transceiver functionalities provided by the individual components into a single or few photonics circuits, thus reducing the number of interconnections required [24]. One challenge in implementing the highest level of PIC is that the passive and active optical components, as well as the large ASIC electronics circuits, are typically fabricated by different processes from different materials. For those three areas, active components such as lasers, modulators, and receivers are best used by III-V materials, e.g., indium phosphide (InP), while passive waveguide-related components such as couplers, splitters, and wavelength multiplexers and demultiplexers normally count on silicon devices. As for suitable ADC-based DSP coherent approaches, they would require a very advanced SiGe or deep-node complementary metal-oxide-semiconductor (CMOS) (40 nm or smaller) process.

6.3 40/100 G SYSTEM ARCHITECTURE AND PRINCIPLES

This section attempts to describe the 40 G/100 G Ethernet structure and its key implementation technology using a multilane distribution (MLD) mechanism, in conjunction with the evolution of OTN for its transport of 40 G/100 GbE [25–27]. Two standard forums, IEEE and the International Telecommunication Union (ITU), have closely cooperated to develop this technology. IEEE 802.3ba is responsible for 40/100 G Ethernet standardization, while ITU-T SG15 is responsible for OTN standardization.

6.3.1 40/100 GbE STANDARDIZATION

The IEEE 802.3ba 40 Gb/s and 100 Gb/s Ethernet Task Force was formed in December 2007. Two new rates of operation for Ethernet were developed: 40 Gb/s for computing and server applications and 100 Gb/s for network aggregation applications. 40 and 100 GbE technology has some similarities to 10 GE in layered system structure, as they are developed from 10 GbE. As shown in Figure 6.12, the common parts of 10, 40, and 100 GbE are the PHY layer and the MAC layer.

Physical layer specifications [25] are tailored to the optimization of each application space, as illustrated in Table 6.1. Four kinds of interfaces are specified for each rate. The maximum transmission distance in the Ethernet standards is 40 km, as the main usage of 100 GbE is networking between networking switches.

One notable feature of 40/100 G Ethernet is multilane transmission. The electrical interfaces are termed 4/10 × 10 Gb/s, the so-called XLAUI or CAUI interfaces for 40 and 100 Gb/s, respectively. It's worthwhile to mention that such an $n \times 10$ G multilane PHY approach is expected to take advantage of the cost reduction by 10 GbE volume.

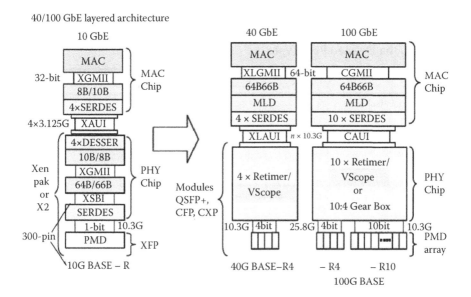

FIGURE 6.12   The architectural partition of a 40 Gb/s and 100 Gb/s Ethernet link into host ASIC IC and optical modules by the XLAUI/CAUI common electrical interfaces.

As multilane transmission is inevitable, the MLD approach has been proposed to facilitate the parallel implementation assisted by simple bit mux/demux. The MLD mechanism is analyzed [28, 29] in Figure 6.13. In the 100 GbE physical coding sublayer (PCS), the data streams are first encoded into 66-bit blocks with the 64 B/66 B coding, and then are distributed among 20 virtual logic lanes through a simple round-robin allocation. In the PMA sublayer, the 20 virtual lanes are bit multiplexed into 10 physical lanes (CAUI interface lanes). Then the CAUI interface lanes are converted in the optical module into 4×25 Gb/s channels to connect to the PMD layer by 10:4 conversion.

6.3.2   ITU-T OTU4

ITU-T, specifically its Study Group 15 Q11, takes care of specifying the OTN evolution for the transport of 40 G/100 GbE being standardized by IEEE. OTN uses a similar multilane mechanism to 100 G Ethernet to achieve 100 G rates.

OTN was standardized in 2001, which provides advanced optical technologies such as optical paths and forward error correction (FEC) suitable for 2.5, 10, and 40 Gb/s WDM systems. The client signals such as Ethernet, SONET/SDH (synchronous digital hierarchy), etc., are wrapped with overhead and FEC bytes. OTN requires transmission over distance greater than 40 km.

The parallel interfaces are defined in an appendix of the G.709 recommendation. The recommendation includes definitions of parallel interfaces for both 40 G and 100 G rates, and within this a new signal was defined for parallel interfaces—the optical channel transport lane (OTL).

TABLE 6.1
IEEE 802.3ba Defined PHY Types

	40 GbE	40 GBASE–	100GbE	100 GBASE–
1m backplane	4 × 10.3125G Optional FEC	KR4		
10m Cu Cable	4 × 10.3125G Parallel Coax Cable	CR4	10 × 10.3125G *Parallel Coax Cable*	*CR10*
100m OM3 MMF	4 × 10.3125G 0.8μm, Ribbon Fiber	SR4	10 × 10.3125G 0.8μm, *Ribbon Fiber*	*SR10*
10km SMF	4 × 10.3125G 1.3μm CWDM	LR4	4 × 25.78125G 1.3μm LAN-WDM	LR4
40km SMF			4 × 25.78125G 1.3μm LAN-WDM	ER4

4-Lane PHY	*10-Lane PHY*

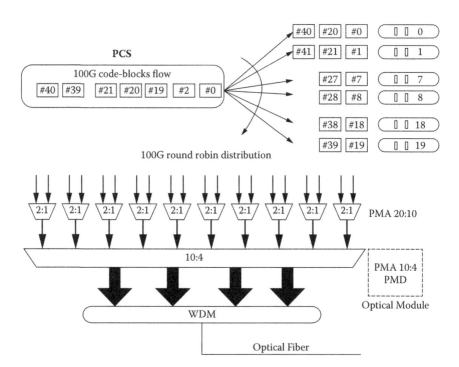

FIGURE 6.13   An illustration of 100 GbE MLD data distribution.

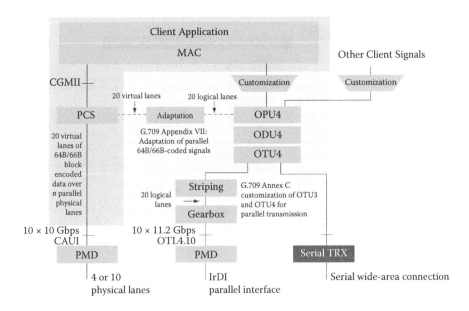

FIGURE 6.14 100 GbE and OTN integration.

To support new 40/100 GbE clients [30, 31], 40 G Ethernet transport will employ the transcoding technique to transport 40 G Ethernet (41.25 Gb/s) over the existing OTU3 of smaller payload capacity (40.15 Gb/s). Equivalently to the XLAUI interface, OTU3 is defined as OTL3.4 in G.709 for the electrical interface of 4 × 10 Gb/s. Similarly, OTU4 specifies the bit rate of 112 Gb/s throughput that supports a 100 GbE stream (103.125 Gb/s) via a standard MLD 10 × 10 Gb/s (CAUI) interface called OTL4.10 for direct connection to SerDes. Alternatively, OTL4.4 denotes an OTU4 mapped over four optical lanes.

Referring to [32] as shown in Figure 6.14, the optics for OTU4 interdomain interface is similar to 100 GBase-LR4 and ER4 applicable to OTU4. The standards support the same high-speed multilane structure for cost-efficient Ethernet over OTN interfaces. This is really essential, as both technologies reuse the same pluggable modules for single-mode interfaces, except the bit rates are slightly different due to the additional OTN or FEC overhead.

6.3.3 100 G TRANSPORT

A rich set of advanced modulation schemes has been explored for 40/100 G, because bandwidth efficiency is crucial. There exist many debates on which modulation scheme is the most appropriate. 40 G deployments are accelerating a move toward DQPSK and away from DPSK, and duobinary. Currently the Optical Internetworking Forum (OIF) [32] has locked in on dual-polarization QPSK (DP-QPSK) following digital coherent receivers for 100 Gb/s transport systems (Figure 6.15). It should be pointed out that by using DP-QPSK, the 4 × 25 G

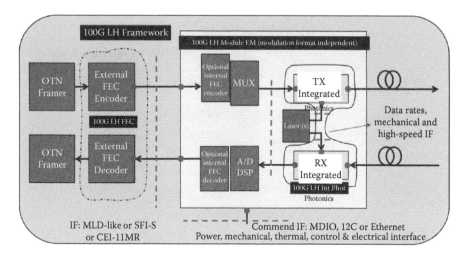

FIGURE 6.15   OIF 100 G DP-QPSK transceiver module architecture.

(28/32 G if ~7/20% FEC overhead is taken into account) electronics, with similarity to the four-lane approach of IEEE 100 GbE, become crucial.

Beyond modulation and coherent detection, system and component vendors are looking to enhanced stronger FEC as an additional tool to improve noise tolerance in 100 Gb/s networks. Approximately 20% FEC overhead is required to achieve the highly demanding performance envisioned for some 100 G backbone DWDM systems [33, 34].

6.4   40/100 G TRANSCEIVER TECHNOLOGIES

Currently the industry is deploying a new generation of optical transceivers for 100 GbE client interconnection and line system transmission targeting data centers and transport applications. This section provides the architecture and overview of the 100 G form factor pluggable MSA for client applications and the OIF 100 G transponder module implementation agreement for long-haul distance DWDM transmission applications.

The introduction of 100 Gb/s is quite different from the introduction of the previous generation. Historically the transition from 1 GbE to 10 GbE brought Ethernet technology to transport networks for the reason that Ethernet has evolved as a ubiquitous protocol. The 40 Gb/s technology has dominated telecommunications for the last 2 years, with substantial modification to the optical transmitter and receivers. One has noticed that physical layer parameters are revolutionizing what is happening today. Now the transmission of 100 G is exhausting all the possibilities one can think of, with more complexity than ever.

Optical transponders (and transceivers) significantly affect the cost structure of the transmission systems, and thus there exists a huge focus on them. Figure 6.16 shows the historical evolution for 10 G telecom modules, and it's expected the 100 G modules will follow a similar trend. Similar to 10 G, all the efforts are directed

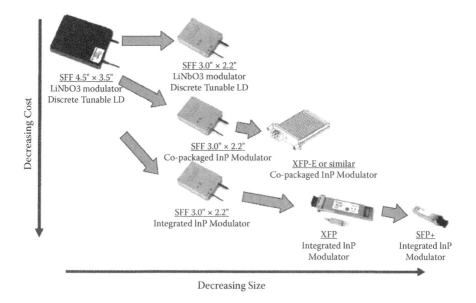

FIGURE 6.16 The historical evolution for 10 G telecom modules (Courtesy of JDSU).

toward the common agreement to build a reasonably priced and compact 100 G technology despite the architectural complexity.

6.4.1 CLIENT MSA MODULES

A series of module form factors have been developed to support the 40/100 GbE optical interfaces, which include CFP, CXP, QSFP+, CFP2, CFP4, QSFP28, etc. Within the Small Form Factor (SFF) Committee, two module form factors have been developed for spatial multiplexing for the anticipated high-volume applications of copper media and multimode fiber. The CFP MSA [33] has defined the module factors that emphasize flexibility at the expense of its large size, which makes it suitable for the highest-density single-mode optical applications. Such module MSAs create a single system configuration that can easily allow system designers to accommodate different optical interface and reach types.

Figure 6.17 shows the roadmap of 100 G pluggable module form factors defined by the CFP MSA. CFP modules are shipping today, supporting multiple pin-map configurations ranging from 3 × 40 G (XLAUI) to OTL4.10. CFP2 modules will double port density, and CFP4 modules will follow a few years later to quadruple port density. CFP2 and CFP4 modules support 4 × 25 G electrical I/O being defined by IEEE.

The CFP block diagram is shown in Figure 6.18 for a generic picture of the common components contained in this first-generation 100 G module. The interface IC for 100 GBase-LR4 and -ER4 applications supports 10:4 gearbox functionality to transition from the CAUI lane mapping of 10 lanes operating at a

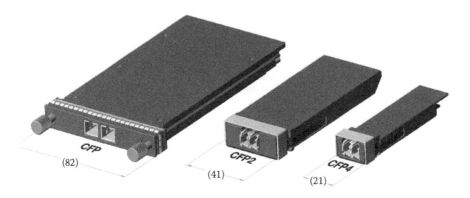

FIGURE 6.17   Various 100 G module form factors defined by the CFP MSA.

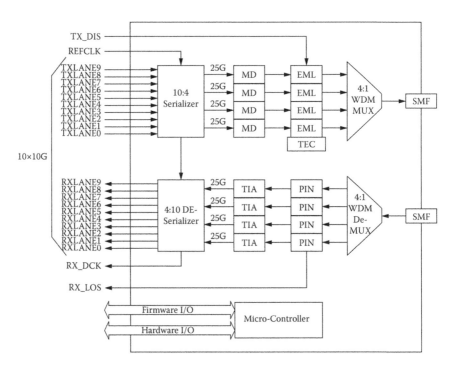

FIGURE 6.18   First-generation 100 Gb/s Ethernet module optics block diagram with $10 \times 10G$ interface.

nominal rate of 10 Gb/s to 4 lanes operating at a nominal rate of 25 Gb/s. The transmitter optical specifications are based upon electroabsorption modulated laser (EML) technology using four LAN-WDM wavelengths at around 1310 nm. The receiver optical specifications can be based upon PIN or APD technology with direct detection.

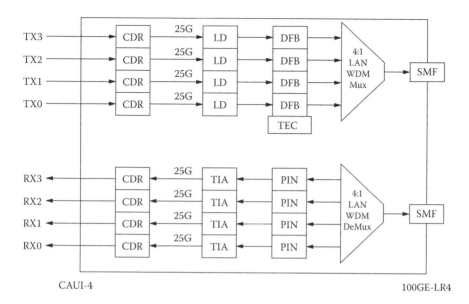

FIGURE 6.19 Second-generation 100 Gb/s Ethernet module optics block diagram using 4 × 25G interface.

The 10 × 10 Gb/s electrical signaling and 10:4 gearbox add complexity, cost, space, and power. For the longer term, gearboxes will be integrated into line card ASIC to build simpler 4 × 25 G electrical I/O. CFP2 and CFP4 are second-generation 100 G form factors using 4 × 25 Gb/s electrical and optical signaling, as shown in Figure 6.19 for an example of 10 GBASE-LR4. The CFP2 and CFP4 platforms enable smaller transceiver form factors and lower power consumption for data center and metro Ethernet applications.

6.4.2 Line-Side MSA Nodules

With the last few years in telecom space, there has been major deployment of the 40 G DWDM system by service providers and carriers in significant volume [35]. Similarly, there are two major drivers for the 40 G deployment: (1) 40 G router-to-router interconnection, which greatly increases the router efficiency, and (2) the economical aggregation of 10 G signals onto 40 G wavelengths in order to increase the spectral efficiency, and quadruple the capacity of the existing DWDM infrastructure.

Due to the technology challenge, 40 G transmission technology was initially implemented in proprietary line card solutions with deployable approaches in DPSK, optical duobinary (ODB), DQPSK, and DP-QPSK. Currently the market has transitioned to MSA transponders from multiple suppliers and volume deployments (Figure 6.20), beginning in 2009. For the client side, the 40 G interface was introduced using the standard very short-reach (VSR) 300-pin-based MSA modules. Nowadays, there is also effort to leverage the CFP form factor with the 40 and 100 GbE applications.

40G Tx/Rx portion of linecard

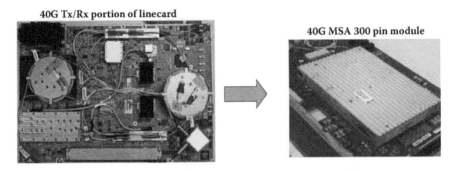

40G MSA 300 pin module

FIGURE 6.20 40 G proprietary line card and MSA modules (Courtesy of Mintera Corp., acquired by Oclaro).

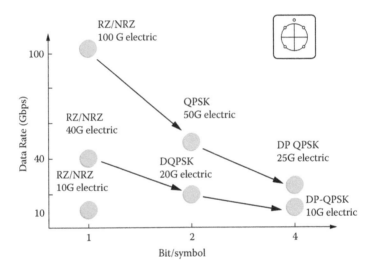

FIGURE 6.21 Various 40 G/100 G available modulation schemes and required electronic rates.

In the past decade, 40 G development has faced significant technology difficulties, and the 40 G market has been quite fragmented with various available modulation schemes (Figure 6.21) to the extent that it even makes the choice for real 40 G deployments quite a challenge. For 100 G development, the industry has learned its lesson from the fragmentation of 40 G markets and the lack of the investment in the industrial standardization. Now with the limited number of modulation format options potentially feasible, 100 G DP-QPSK MSA modules have been adopted through industry forums such as OIF.

The block schematics of a 40/100 G DP-QPSK transmitter is illustrated in Figure 6.22. For the transmitter design, the center component is two QPSK modulators for each polarization, which consists of two nested Mach-Zehnder modulators

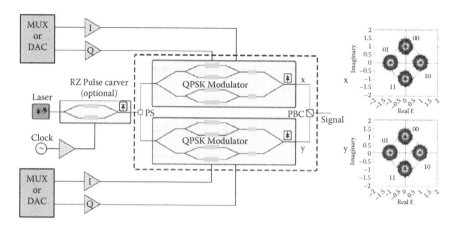

FIGURE 6.22 40 G/100 G DP-QPSK transmitter block schematics.

with bias control, a 90° phase shifter with phase control, and a monitoring photo-diode (MPD) for output optical power monitoring. A continuous wave (CW) laser source output is split by a beam splitter (BS) into two components (X, Y), with each component independently modulated by a QPSK modulator. Modulator drivers and associated electronics for in-phase (I) and quadrature (Q) high-speed data encoding, I/Q bias control, phase control, and optical power control are external to the QPSK modulator.

The modulated signals are recombined by a beam combiner (BC) with their polarizations orthogonal to each other and transmitted on an output optical fiber. The QPSK modulator is specified to provide a minimum bandwidth of 23 GHz for supporting applications with nominal symbol rates of up to 32 Gbaud.

The block diagram of a DP-QPSK coherent receiver is illustrated in Figure 6.23. The integrated dual-polarization intradyne coherent receiver is the critical optics component with detailed specification of optical, electrical, and electro-optical properties. For the coherent receiver design, a polarization beam splitter (PBS) is used to split the received optical signal (from a DP-QPSK transmitter) into two components (X, Y) with orthogonal polarizations. These polarized optical signals are mixed with a local oscillator laser source at a frequency near that of the received optical signal, generating mixing products at the difference frequency. The resultant analog products are down-converted, detected electronically, linearly amplified, digitized in high-speed ADCs, and then passed to a DSP ASIC chipset. The DSP ASIC chipset includes equalization and FEC functionality for propagation impairment compensation and high-speed electrical SerDes interface functionality.

Optical transceivers of 40 and 100 Gbit/s use digital coherent technologies, which realize ultra-long-haul transmission over 2000 km, as well as high capacity with a single carrier. These transceivers are designed for DWDM optical systems and high-speed routers by applying the standard MSA form factor for each bit rate.

It's worthwhile to mention that the transceiver interface IC may optionally implement hard- or soft-decision forward error correction (HD- or SD-FEC)

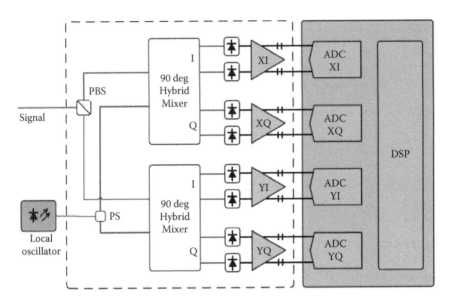

FIGURE 6.23 100 G DP-QPSK coherent receiver functional architecture at 32 Gbaud.

encoder/decoder functionality for improving OSNR performance in 100 G transmission. SD-FEC coding divides the signal level space into finer divisions for discriminating between 1 or 0 bit, and thus offers the potential of higher net coding gain than HD-FEC, which simply uses a single level for 1/0-bit discrimination. However, SD-FEC implementation requires increased coding processing bandwidth and performance; thus, there is a trade-off between net coding gain improvement and implementation complexity/performance penalties.

6.5   ENABLING HIGH-SPEED PHYSICAL LAYER ELECTRONICS

For the first time in networking, the system design is primarily limited by single-lane electronic process speed for the sake of power, cost, and die size. Along with the increase of data transmission speed, signal distortion, and thus ISI, signaling and equalization become much more important. All 100 GbE (4 × 25 G), 40 G DQPSK, and 100 G DP-QPSK take advantage of the parallel implementation or alternative modulation schemes, and hence leverage the ~28 G IC chipsets in the range of 25–32 Gb/s [36, 37]. In the physical layer, the powerful eFEC codes are also one of the critical tools in consideration. Another bottleneck is the need of fast ADC (or DSP) devices requiring 2 symbols/bit for the case of coherent detection.

In addition to ITU-T-specified FEC of 7% overhead, powerful FEC codes of approximately 20% overhead are needed for 100 G DP-QPSK. Both hard- and soft-decision FECs are demonstrated for coding gain of over 10 dB [34]. Hard-decision FEC has proved the most feasible to implement for 40 and 100 Gb/s in the form of either a field-programmable gate array (FPGA) code or application-specific integrated circuit (ASIC) implementation. However, the implementation of soft-decision

FEC, while meeting power, size, and cost targets, still represents a significant technical challenge and requires substantial innovation.

6.5.1   25 G Signaling and EDC

The current 40/100 GbE is initially implemented by electrical interfaces that utilize a 10 lanes by the 10 Gb/s signaling rate. Ultimately, power and size limitations associated with such a wide interface will drive the need for a faster and narrower interface. Recognizing this need, OIF's Physical and Link Layer (PLL) Working Group has just recently initiated the Common Electrical Interface 28 Gb/s project (CEI-28 G-VSR) [36, 37], which includes electrical specifications for 28 Gb/s signaling for chip-to-module applications over very short reaches up to 200 mm. As shown in Figure 6.24, this work will enable narrower four-lane interfaces for 100 Gb/s applications, such as 100 GbE and OTU4, which will enable smaller package sizes, lower pin count components, connectors, and optical modules, lower power dissipation, and clockless interfaces ideal for data center, server, and core router applications.

In this subsection, we discuss two receiver architectures for equalizing 28 Gb/s NRZ (non-return-to-zero) signals over the CEI-28 G-VSR proposed channel. Both equalizer architectures, also referred to as electronic dispersion compensation (EDC) architectures, are consequently compared from the perspective of cost, power, complexity, and performance.

6.5.1.1   28 G-VSR Channel Model

In order to support 100 Gb/s data rates robust over narrower interfaces, low-power equalization and clock and data recovery (CDR) functionality must be included in the module, as illustrated in Figure 6.24. The optimal design of such functionality is contingent on thorough analyses of the channel. The channel model [38, 39] adopted here consists of a one-connector PCB physical setup with a reach of 102 mm to the host side and a reach of 31 mm to the optics module side. This particular channel's characteristics are summarized in Figures 6.25 and 6.26. Figure 6.25 illustrates the channel's frequency-domain insertion loss (IL) and return loss (RL). Figure 6.26 illustrates the channel's time-domain impulse response. In particular, two quantities

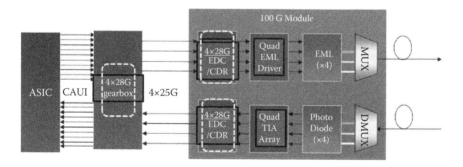

FIGURE 6.24   Next generation implementations for 100 GBase -LR4 and -ER4.

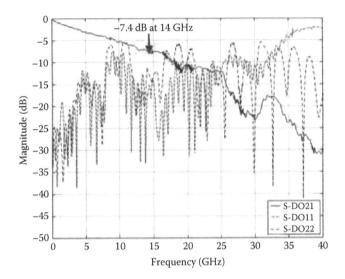

FIGURE 6.25   A 133-mm CEI-28 G-VSR channel insertion loss and return loss.

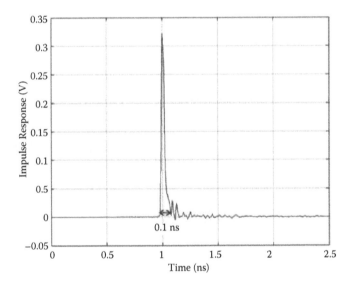

FIGURE 6.26   A 133-mm CEI-VSR channel impulse response.

of interest are the insertion loss magnitude of –7.4 dB at the Nyquist frequency of 14 GHz and the main impulse response width of 0.1 ns.

6.5.1.2   Signaling and Equalization

The channel's low-loss property (–7.4 dB) and its relatively short intersymbol interference (ISI) time span (0.1 ns) allow the use of NRZ signaling and low-complexity equalization. This is critical to achieve the low-cost/power implementation targets of EDC/CDR devices, especially when running at such single baud rates as high as 28 Gb/s.

Low-complexity and low-power equalization are achievable either with the use of continuous-time linear equalizers (CTLEs) [40] at the receiver coupled with additional equalization at the transmitter in the form of preemphasis, or with the use of adaptive linear transversal feedforward equalizers (FFEs) at the receiver [41]. In the next paragraphs and sections, we investigate and compare the performance of both approaches.

CTLE filters typically equalize a given channel by providing high-frequency boost in order to achieve a flat-channel frequency response. CTLE transfer functions are determined by proper location of the poles and zeros. Depending on the channel characteristics, typical CTLE transfer functions will contain either one zero and two poles corresponding to a first-order channel model, or two zeros and three poles corresponding to a second-order channel model by Equations (6.1) and (6.2), as given below:

$$H_1(s) = \frac{K_1(s + z_1)}{(s + p_1)(s + p_2)} \tag{6.1}$$

$$H_2(s) = \frac{K_2(s + z_1)(s + z_2)}{(s + p_1)(s + p_2)(s + p_3)} \tag{6.2}$$

CTLE zero location is usually performed by fitting first-order (one pole) or second-order (two poles) low-pass models through the channel frequency response. This is typically an optimization process by which the CTLE zero frequency values are varied until the best model fit to the channel is achieved. Because of the inverse relationship between channel model and CTLE, the poles of the channel model correspond to the zeros of the CTLE. CTLE pole location is usually a function of overall filter peaking gain and the data rates. This is further illustrated in Figures 6.27 and 6.28. Figure 6.27 shows the channel frequency response, the first-order fitted channel model, the CTLE filter with a peaking gain of around 6.5 dB, and the resulting equalized channel. Figure 6.28 depicts the channel frequency response, the second-order fitted channel model, the CTLE filter with a peaking gain of around 6.5 dB, and the resultant equalized channel.

The adaptive linear transversal feedforward equalizer is a three-tap FFE filter at the receiver whose coefficients are updated according to the least mean square (LMS) algorithm [42]. In the following section, we present eye diagram simulations to illustrate the performance of both equalization approaches. The eye diagrams were generated using a time-domain-based simulator.

6.5.1.3 Simulation Results

The time-domain simulations consist of a 28 Gb/s transmitter generating PRBS $2^{31} - 1$ NRZ patterns with added random jitter and periodic jitter. The channel impairments consist of a package model at both transmitter and receiver ends, the 133 mm PCB channel with its associated impairments of IL, RL, and crosstalk,

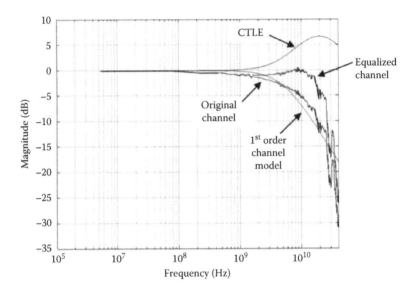

FIGURE 6.27   A CTLE filter with Z_1 = 4.86 Hz, p_1 = 19.36 Hz, p_2 = 23.26 Hz, and peaking gain = 6.5 dB.

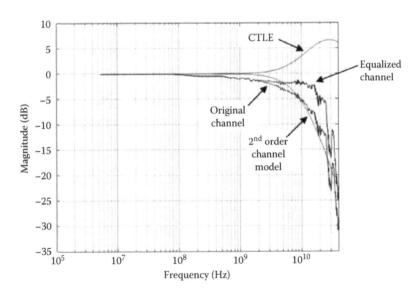

FIGURE 6.28   A CTLE filter with Z_1 = 4.86 Hz, Z_2 = 9.76 Hz, p_1 = 20.76 Hz, p_2 = 24.86 Hz, p_3 = 28.96 Hz, and peaking gain = 6.5 dB.

and additive white Gaussian noise (AWGN). The crosstalk waveforms are generated using aggressor transmitters similar to the reference transmitter. Two sources of near-end crosstalk (NEXT) and two sources of far-end crosstalk (FEXT) were included in the simulations. The receiver implementation includes both equalization techniques as described above.

The nonideal transmitted eye pattern of the package output is illustrated in Figure 6.29(a), where the effects of total jitter, package IL, and package RL are highlighted. The received eye diagram with no equalization and no transmitter preemphasis is illustrated in Figure 6.29(b), which indicates a pretty closed eye. For comparison, the received eye diagram with no equalization and 3.5 dB transmitter preemphasis is illustrated in Figure 6.29(c). Transmitter preemphasis is accomplished via a two-tap finite impulse response (FIR) filter with programmable preemphasis levels. The received eye diagram equalized with the first-order channel model CTLE of Figure 6.27 is illustrated in Figure 6.29(d). Accordingly, the received eye diagram equalized with the second-order channel model CTLE of Figure 6.28 is illustrated in Figure 6.29(e). The received eye diagram equalized with a three-tap FFE LMS is illustrated in Figure 6.29(f).

Both first-order and second-order CTLE equalizers provide similar vertical eye openings; however, the second-order model appears to have wider horizontal eye openings and less transient behavior. Further optimization with CTLE pole/zero location and peaking levels may provide additional performance gains.

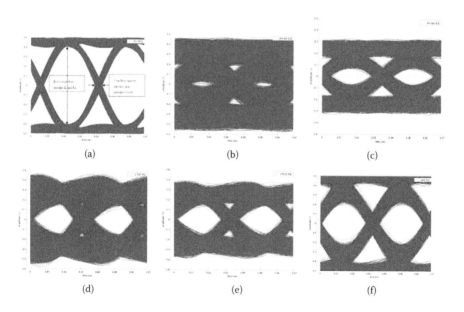

(a)   (b)   (c)

(d)   (e)   (f)

FIGURE 6.29 (a) Transmitted eye pattern at package output with no Tx pre-emphasis at $1V_{pp}$; (b) Received eye diagram without pre re-emphasis and no equalization; (c) Received eye diagram with 3.5 dB Tx pre-emphasis but no equalization; (d) Received eye diagram with 1st-order CTLE of Fig. 6.27; (e) Received eye diagram with 2nd-order CTLE of Fig. 6.28; (f) Received eye diagram with 3-tap FFE LMS.

Moreover, the three-tap FFE LMS equalizer is superior to its CTLE counterparts. Further optimization with the LMS convergence parameters may provide additional performance margins.

Additional performance improvements to the CTLE architecture could be achieved by implementing transmitter preemphasis with a three-tap FIR filter. This approach provides precursor and postcursor ISI cancellation capabilities. However, calculation of the optimal FIR coefficients may require additional system complexity, and consequently, the final architectural solution may become very comparable to the FFE LMS architecture.

6.5.2 CDR AND SERDES

The evolution of the 100 GbE transceiver interfaces is expected to follow a path similar to that of the prior 10 GbE technologies, which evolved from large, power-hungry, and niche market units to compact, lower-power, and mainstream products. The first critical step on this path will be to migrate the 100 GbE transceiver's high-speed SerDes function from its present implementation in exotic SiGe processes to low-power CMOS designs that can be economically fabricated using mainstream commercial CMOS processes. The second step in this evolution is to improve the signal integrity of the SerDes to allow it to be moved out of the 100 GbE transceiver and onto the line card, to reduce the power consumption and the size of the transceiver.

Developing a low-power CMOS SerDes transceiver capable of supporting a highly reliable 25 Gbps data stream is one of the keys to enabling higher port density [43]. The cost and power savings over an equivalent SiGe circuit are due to CMOS's simpler, inherently compact structures, which can be easily scaled downward as semiconductor processes improve. Line card and optical module manufacturers using CMOS products will benefit from the large community of competing foundries that engage in aggressive pricing strategies and rapid adoption of ever-smaller process nodes that deliver successively lower per-chip costs, reduced operating voltages, and lower power consumption.

An example of the block diagram of the 4 × 25 G 10:4 SerDes (or gearbox) is shown [44, 45] in Figure 6.30. On the transmission direction of the gearbox, ten 10.3125 Gb/s physical lanes are converted into four 25.78125 Gb/s physical lanes. On the receiving direction, on the other hand, four 25.78125 Gb/s physical lanes are converted into ten 10.3125 Gb/s physical lanes.

Because the 10:4 multiplexer does not have an integer ratio, implementation is difficult. Therefore, the 10 physical lanes are first converted into 20 PCS lanes, each at the serial speed of 5.15625 Gb/s. Then, the 20 PCS lanes are converted to four physical lanes. That is, there are four parallel 5:1 bit multiplexers (transmitting direction) and 1:5 bit demultiplexers (receiving direction). These implementations of multiplexers and demultiplexers may result in switching of the orders in which the PCS lanes appear at the receiving side of the PCS. But this is not an issue, as the PCS lanes will be reordered at the receiving side of the PCS.

High-speed, low-power CMOS architectures are needed to provide complete solutions for physical layer ICs inside the transceiver module and on the line card to achieve these two critical steps. Once achieved, they will enable production of

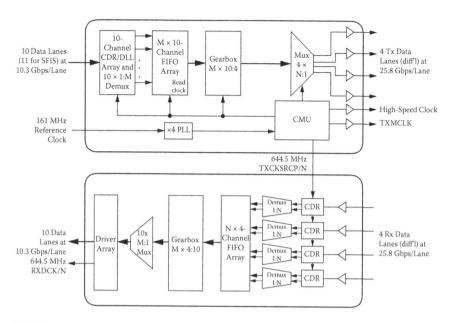

FIGURE 6.30   An exemplary block diagram of 100 G SerDes (10:4 GearBox).

next-generation 100 GbE systems that offer up to 10× higher port density while reducing transceiver power consumption by more than 50%.

6.5.3 COHERENT DETECTION AND DSP

There are many considerations for implementing the ADC and DSP in coherent detection [47, 48]. A DP-QPSK coherent receiver needs four ADC channels since there are two optical polarizations and each needs two ADCs to digitize an I/Q signal. To achieve a 100 Gbps net line rate, a baud rate of at least 28 Gbaud/s is used to allow for FEC overhead, which needs 56 GSa/s ADCs. The system SNR requirements mean that 6- to 8-bit resolution is typically required to allow some margin for added noise and distortion; so for four ADCs, the output data rate to the DSP is 1.3 Tb/s, or 1.8 Tb/s if 8-bit resolution is used to allow more noise margin or digital automatic gain control (AGC) and equalization after the ADC.

Normally ADCs should be integrated with the DSP; otherwise, this huge amount of data has to be transmitted between chips, which is difficult to implement (very large number of channels with high data rate). A 100 G receiver DSP, which performs functions such as equalization, chromatic dispersion compensation, and data recovery, needs on the order of 50 M gates, which mandates the use of CMOS.

The system power requirement for a complete coherent receiver is only a few tens of watts; since a 40 G ADC/DSP chip in 90 nm already dissipates more than 20 W, geometries of 65 nm or smaller, as well as power-efficient design techniques, are needed for a 100 G receiver. This implies that the ADC should also use CMOS, although this means the design is extremely challenging. A single-chip solution is really the only viable way forward, especially in order to take advantage of future

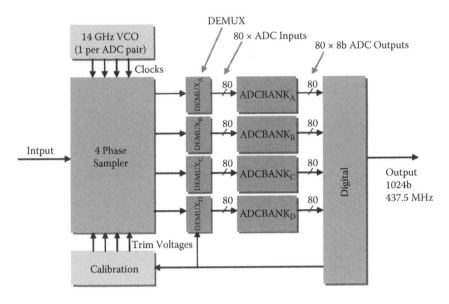

FIGURE 6.31   Block diagram of 56 GSa/s ADC/DSP for 100 G DP-QPSK [47].

CMOS technology improvements—though this does assume that the ADC performance scales similarly to digital circuits, which may not be true for conventional
ADCs. Even if a multichip solution can be built—for example, using SiGe ADCs
together with a CMOS DSP in a multichip module (MCM)—then not only will the
overall power be higher, but also the production cost will be greater and the yield of
such a complex solution will inevitably be lower. This also does not give a good roadmap toward even lower-power and lower-cost solutions for short-haul and beyond
100 G.

Designing a ~56 GSa/s 6–8 b ADC in any technology presents major difficulties.
The challenge is even more difficult in this case, because available power for the
ADC + DSP is limited by both supply capability and thermal dissipation. A reasonable target is 10 W or less for a complete four-channel ADC, which means little more
than 2 W per ADC cell. To overcome these challenges, one solution is to use a new
sampler/demultiplexer architecture [47, 48] that gives the linearity, noise, and bandwidth required without needing extremely short-channel (40 nm or below) transistors; allows simple calibration of amplitude and timing errors during operation; and
dissipates <0.5 W. Instead of a conventional S/H using analog switches and sampling
capacitors, this circuit generates controlled-shape constant-area (charge) sampling
pulses that are then demultiplexed to drive a large array of 8 b SAR ADCs (320 ×
175 Ms/s) (Figure 6.31). Using SAR ADCs instead of full flash means that increasing
resolution from 6 b to 8 b carries only a small penalty in power and area.

6.5.4 FORWARD ERROR CORRECTION

Forward error correction (FEC) has been integrated as a necessary adjunct for
10 Gbit Ethernet networks, and is assumed to be a necessity for any system operating

above that speed, particularly in the next-generation Ethernet speeds of 40 and 100 Gb/s. FEC relies on a standardized way of sending redundant data in the same channel as a message, providing a means by which the receiver can recover information if the channel is corrupted.

The concept underlying FEC was introduced by Shannon [21] in a paper that laid the foundations of modern digital communications. In this 1948 paper, best known for its expression deriving the capacity boundary of a noisy channel, Shannon proved the remarkable information capacity theorem for feeding forward to the receiver error data based on a comparison of the transmitted and received signals in a noisy channel. Shannon's model postulated an external observer suitably placed to be able to make the comparison. However, these results cannot be directly applied to fiber optic communications systems [48]. In practice, FEC is implemented by adding to the transmitted signal data a parity code that enables the receiver to detect and evaluate the data errors due to noise in the transmission channel, thereby providing a mechanism for their correction without requiring Shannon's external observer.

6.5.4.1 Deployment History of FEC

Figure 6.32 plots the progress in FECs for optical communication systems over the last two decades [49]. The vertical axis shows the product of linear net coding gain, defined in terms of a post-FEC BER of 1×10^{-15} and bit rate in Gb/s (NCG = net coding gain: gross coding gain is dB minus the bit rate increase in dB due to the redundant overhead). This value is an analogy of gain-bandwidth product in electrical amplifiers. The three sets of data points represent the different generations

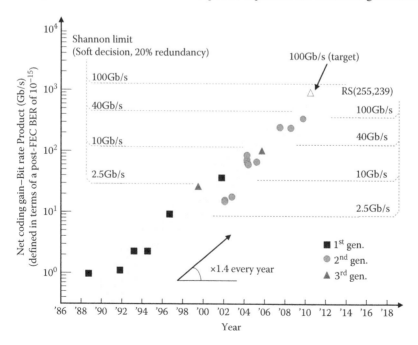

FIGURE 6.32 Progress in FECs for optical communications.

of FEC: first-generation FEC using Reed-Solomon (RS) linear block codes represented by RS (255, 239), second-generation FEC using concatenated sets of codes, and third-generation FEC based on soft decision and iterative decoding. All of these points represent FECs that are not just simulated, but have actually been subjected to experiments or deployments in live physical circuits.

One of the first published practical FEC experiments in optical fiber communications was reported by Grover in 1988, by inserting a shortened Hamming code onto 216 data code bits, resulting in a (224, 216) encoded signal [50]. The 3.7% redundancy, giving a 0.96 code rate, was added to the low-speed tributaries of a 565 Mb/s signal, yielding an FEC coding gain of about 2.5 dB at an output BER of 1×10^{-13}. The first fully fledged deployment of FEC for optical transmission was in submarine systems developed in the 1990s. Well-known block codes, RS codes, were demonstrated in submarine transmission cable systems in that time frame [51]. The RS (255, 239) code, in particular, then came to be implemented in a broad range of long-haul systems as recommended by ITU-T G.975 and G.709. We refer to this era's FECs, based on hard-decision decoding (single quantization level bit sampling), as first-generation FEC.

As wavelength-division multiplexing (WDM) matured in the 1990s, system designers started to seek more powerful FEC schemes than the first-generation FEC. Since the end of the 1990s, several types of powerful FEC have been developed based on concatenating mathematically orthogonal codes [52], e.g., RS (239, 223) + RS (255, 239) for superior error correction (the outer code is used to mop up any errors left uncorrected by application of the inner code). We refer to this class of concatenated codes with hard-decision decoding as second-generation FEC. Interleaving and iterative decoding techniques have been used together with the concatenation to obtain improved error correction performance. Several types of concatenated codes are listed in ITU-T G.975.1, e.g., a pair of orthogonally interleaved BCH codes having an NCG of 10.06 dB at a post-FEC BER of 1×10^{-15}. A number of second-generation FEC developments and transmission demonstrations have been reported to date. An optical transport network (OTN)-compatible enhanced FEC (eFEC) large scale integrated circuit (LSIC) [53] having an NCG of 8.5 dB at post-FEC BER of 1×10^{-15}, in reference to the generic RS (255, 239) codes ratified by ITU-T, is widely used for 10 and 40 Gb/s WDM systems.

System designers have shifted their interest to intensively search for even more powerful FEC schemes having an NCG of over 10 dB. We refer to this class of FEC as third-generation FEC. The candidates for this application are soft-decision and iterative decoding. The first demonstration of a soft-decision-based FEC in optical communication systems was a concatenation of an RS code and a Viterbi convolutional code in 1999 [54]. The two codes together delivered an NCG of 10.3 dB at 2.5 Gb/s, but the redundant overhead was as high as 113%; it required 5 Gb/s to transmit 2.5 Gb/s of signal data. The second demonstration was of a block turbo code with 3-bit soft decision having an NCG of 10.7 dB (at post-FEC BER of 1×10^{-15}) and an overhead of barely 24.6% [55]. These progressive challenges, however, were not fully integrated into LSI, with only a part of the channel integrated for soft-decision decoding. In 2006, a fully integrated block turbo code scheme for 12.4 Gb/s was reported for the first time [56]. Recently, low-density parity-check (LDPC) codes [57, 58] are awakening promise of a third-generation FEC able to approach the Shannon limit.

Looking again at Figure 6.32, a clear trend can be seen in that an improvement of 1.4 times has been achieved on average every year. This improvement has been achieved not only by FEC algorithm improvements going from first generation to third generation, but by LSI technology evolution. The open triangle at the top shows the research target for 100 Gb/s FEC: the reach of >10 dB NCG and a 100 Gb/s bit rate. However, the realization of an NCG of >10 dB still represents a challenge, because the FEC redundancy in high-speed optical communications is limited by, among others, the availability of high-speed analog devices, the associated optical components, and the complexity of the digital circuitry. The industry consensus is that the maximum practical redundancy is currently not beyond 20% for 100 Gb/s digital coherent systems.

6.5.4.2 Hard-Decision-Based FEC

Traditionally, block FEC codes have been implemented using hard-decision algorithms, in which a hard-decision decoder makes firm decisions for every input and output as to whether the signal corresponds to a 1 or 0. Unlike soft-decision algorithms, hard-decision algorithms will not indicate the reliability of a decision. Hard-decision implementations are the current practical alternative for 40 and 100 Gb/s transmissions.

Recent work in improving continuously interleaving versions of traditional Bose-Chaudhuri-Hocquenghem (CI-BCH) cyclic error-correcting codes have demonstrated the achievement of an NCG of 9.35 dB within the standard 7% OTN overhead rate [59]. An early demonstration of an enhanced FEC (eFEC) using CI-BCH (1020, 988) has demonstrated that a 2.5 Gb/s implementation in a field-programmable gate array (FPGA) can be configured for 6.7% overhead or higher overhead rates. It can detect up to four errors, and correct three. Simulations have shown that with greater overhead of up to 20%, an NCG of more than 10 dB can be achieved.

The ITU-T G.975.1 has documented several enhanced FEC codes, with CI-BCH specified as G.975.1-I.9. In recent months, partially as an expansion of the work within IEEE on 802.3ba higher-speed Ethernet, ITU-T SG15 has initiated a new effort in conjunction with IEEE to create a new transmission rate and signal format for transport of a 100 Gb/s Ethernet signal. ITU-T SG15 has identified a new level of optical transport network hierarchy for this purpose, called the optical channel transport unit level 4 (OTU4) signal, a signal that also will support multiplexing of lower-rate OTN signals. The OTN standards allow the use of proprietary FEC codes within the OTU overhead.

In joint discussions with OIF, ITU-T considered the possibility of standardizing such FEC codes, but elected to reserve the standardization of the generic RS (255, 239) FEC at 100 Gb/s only for the intradomain interfaces (IrDIs) between two or more service providers, typically short-reach connections. Studies have continued within ITU-T on higher-gain FEC codes for metropolitan applications, but ITU-T has deferred study on such standardization efforts, while OIF is concentrating on the development of a coherent receiver module, based on digital signal processing (DSP) techniques, for long-haul applications.

Many carriers have chosen to use existing 10 Gb/s OTU2 networks for transmission of 40 Gb/s OTU3 traffic, utilizing a combination of advanced modulation (DPSK or DQPSK), electronic dispersion compensation (EDC), and higher-gain FEC, in an effort to avoid reengineering fiber plant for higher-rate signals. CI-BCH can be particularly useful in this regard because of its reduced encode-decode latency—the latencies encountered by some service providers in OTN hierarchies had earlier led to calls by some service providers for disabling FEC altogether.

Several companies such as Vitesse previously standardized a 10 Gb/s design for G.709 and eFEC, implemented in single, dual, and quad versions [59]. The company's FEC solutions typically operate together with discrete EDC and SerDes chips or now integrate EDC functionality to optimize the footprint. The dual and quad FEC devices are based on a 65 nm CMOS process technology.

The CI-BCH FEC code is constructed from continuously interleaved BCH (1020, 988) codewords. A complete description of the CI-BCH FEC code is provided in [59]. Computed FEC bytes are stored in the standard G.709 OTN FEC byte locations. This enables the received signal to be processed with or without decoding FEC bytes. CI-BCH FEC tolerates a burst of up to 1500 consecutive errors, and theoretically delivers >9.35 dB NCG at an output BER of 1×10^{-15} and a decoding latency of approximately 1 Mbits, as shown in Figure 6.33. Note that this represents a better coding gain than any G.975.1 FEC codes at a standard 7% overhead.

The use of concatenated, interleaved BCH codes simplifies FEC decoding, as an algebraic equation solver can be applied, rather than more complex Chien search or matrix inversion techniques. CI-BCH has an extremely low flaring floor, eliminating the need for flaring correction in most applications. Finally, the use of continuously interleaved, identically constructed BCH codewords enables the decoder to decode fewer interleaved codewords, reducing latency at the expense of reduced coding gain.

This code architecture provides double coverage of all transmitted bits by a triple error-correcting BCH, allowing high-gain iterative decoding to proceed. It has only a single type of codeword that acts in a manner similar to both the inner and the outer

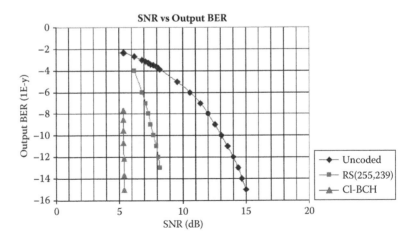

FIGURE 6.33 Output BER vs. SNR for CI-BCH and RS (255, 239) FEC codes.

code of other high-gain codes. It is this aspect of the code that keeps the latency of this code similar to the other high-gain proposals. Extensive analysis in the simulation of 1×10^{17} bits transmitted at high-input error rates and density evolution techniques outlined by Richardson and Urbanke [60] have shown that this type of BCH (1020, 988) base code and interleaver achieves nearly optimal performance for 7% overhead hard-decision decoded high-gain codes at 1-megabit latencies.

Basing the code on a triple error-correcting BCH allows the error locator to be strictly algebraic. This allows very efficient and high-gain iterative decoding to proceed, and thus a low-latency decoder to be designed, because no Chien search is required. All code parity bits are a function of only previously transmitted bits. The consequence of this is that no matrix inversion is required to achieve double coverage of all transmitted bits. Table 6.2 indicates theoretical NCG performance improvements for the proposed FEC code. The examples shown in the table are for a three-error-correcting BCH for standard 6.7% overhead, and for higher overhead, a coding gain of 10.3 dB at 20% is achievable [61]. When a four-error-correcting CI-BCH-4 is considered, latency is 8 Mbits, and coding gain varies from 9.55 dB at an overhead of 6.7%, to a coding gain of 10.5 dB at 20% overhead.

Each data point in Table 6.2 represents between 1×10^{-14} and 2×10^{-16} simulated transmission bits. Errors were distributed in an uncorrelated random fashion. These results implement the errors remaining after 1 megabit of latency. Two sweeps through the syndromes were performed upon the reception of each G.709 subframe. Simulations were also run with small-range-correlated burstiness with no difference in performance. In the absence of other errors, this code can withstand a maximum consecutive burst of errors of 1500 bits.

To verify actual CI-BCH FEC performance, the CI-BCH code was first implemented in an FPGA operating at 2.5 Gb/s OTU1 line rates. Both electrical and optical performance testing were completed. In the electrical test setup, an error generator was included in the FPGA after the FEC encoder, which injected Poisson-distributed errors expected due to additive white Gaussian noise (AWGN). Subsequently, the 2.5 Gb/s FEC-encoded OTU1 signal without error insertion was transmitted into an optical test setup that included fiber, an optical attenuator, and an optical noise source. Uncorrected input BER was measured before FEC decoding. Corrected output BER

TABLE 6.2

Theoretical NCG for Proposed Continuously Interleaved BCH FEC

Channel Input Error Rate	Corrected Output Error Rate	NCG
4.80×10^{-3}	2×10^{-8}	6.25 dB
4.70×10^{-3}	3×10^{-10}	7.23 dB
4.65×10^{-3}	2×10^{-11}	7.81 dB
4.60×10^{-3}	7×10^{-13}	8.40 dB
4.55×10^{-3}	2×10^{-14}	8.99 dB
$\leq 4.5 \times 10^{-3}$	$< 10^{-15}$	9.38 dB

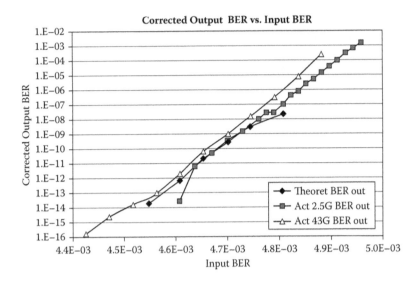

FIGURE 6.34   Actual performance of CI-BCH at 2.5 Gb/s OTU1 and 43 Gb/s OTU3.

was measured after CI-BCH decoding. Corrected 2.5 Gb/s output BER vs. uncorrected input BER was plotted, as shown in Figure 6.34 [61].

Following successful 2.5 Gb/s CI-BCH performance, the encoder and decoder were subsequently implemented in a single FPGA to support 43 Gb/s OTU3 operation. The same random error generator was also implemented in the FPGA. The CI-BCH-encoded OTU3 signal with errors injected is looped back to the CI-BCH FEC decoder. Once again, uncorrected input BER was measured before FEC decoding, and corrected output BER was measured after CI-BCH decoding, with the 43 Gb/s results plotted as shown in Figure 6.34.

6.5.4.3 Soft-Decision-Based FEC

Soft-decision detection is performed by setting $2^N - 1$ decision thresholds, where N is the number of quantization bits. Figure 6.35 illustrates a typical soft-decision structure for a QPSK constellation. The received signal is digitally converted to M-bit steps by a network of analog-to-digital converter (ADC) thresholds. The thresholds sandwiched between the two signal states are used for the soft-decision decoding. In the case of $N = 3$ as depicted in Figure 6.35, the two possible received signals, 1 and 0, lie in regions represented by a binary vector ranging from [011] to [111]. The left-most bit is the hard-decision digit, and the other two digits are information bits indicating the probability of 1 and 0.

In wireless systems, an N of 8 is easily generated by an ADC operating at tens to hundreds of megasamples per second. Since ADC construction becomes more difficult for higher-speed signals such as 10 to 60 gigasamples per second in the case of optical communications, very few trials have been made of soft-decision detection. Conveniently, a recently appearing digital coherent receiver incorporates an ADC at its front end for demodulating multilevel coded signals. This suddenly

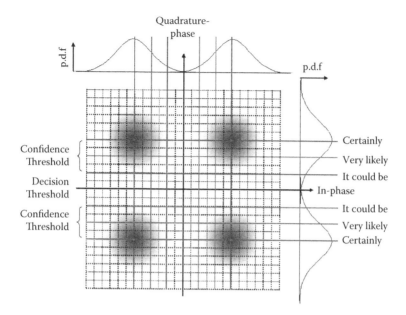

FIGURE 6.35   Typical 3-bit soft decision for multi-level coded format.

makes it much easier to realize soft-decision decoding, even though the symbol rate is much higher than in wireless systems.

The coding gain can in principle be improved by $\pi/2$ by using soft-decision decoding with infinite quantization bits and redundancy. This is approximately a 2 dB difference.

6.5.4.3.1   LDPC Codes

For 100 Gb/s optical communications, one potential candidate for strong soft-decision-based FEC is an LDPC code, a linear code defined by the sparse parity-check matrix invented by Gallager [57]. The LDPC code is expected to exhibit superior error-correcting performance close to the Shannon limit. This technique was ignored for a long time until it was rediscovered by MacKay [62], leading to further intensive research for wireless communication systems. Later, studies into applying LDPC to optical communications were instituted. The expected merits of LDPC codes are not confined to such high error correction capability, but also fit them for parallelization to reduce circuit complexity. In an LSI for high-speed optical communications, ease of parallelized signal processing is essential for practical implementation. Among the various coding schemes, it is commonly believed that LDPC is the most suitable code for parallelization.

Although the LDPC codes show superior potential as indicated above, there is a tough problem to solve: "error floor" [63, 64], often observed in the measured post-FEC BER. Like many iterative codes, the performance curve steepens with rising SNR, up to a point where it reaches an error floor and the curve slope then flattens. Even though an FEC may show superior performance at a post-FEC BER of 1×10^{-7}, if there is an error floor at 1×10^{-8}, such an FEC is unusable because a minimum post-FEC BER of at least 1×10^{-12}, and preferably 1×10^{-15}, is generally required in optical transport

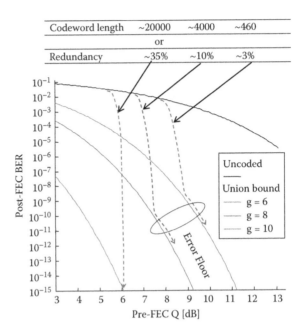

FIGURE 6.36 Typical BER performance of LDPC codes.

systems. For practical LDPC implementations, especially at low BER, error floors can be quite hard to measure due to long simulation or emulation times.

Figure 6.36 shows the typical BER performance for regular LDPC codes having only columns of weight four. The solid lines show the union bounds, a union bound being the approximation of post-FEC BER with maximum likelihood decoding at the high Q limit. The union bounds are affected by the girth g, which is the length of the short loop in the parity-check matrix. In the case of message-passing decoding, the girth dominates the error correction performance in the low post-FEC region. In order to reduce the error floor, larger g (~10) would be necessary. It is worthwhile to mention that Figure 6.36 shows a conceptual BER performance of LDPC codes. By designing LDPC codes properly, better BER performances can be achieved [58].

There are two ways to increase the girth. One is increasing the codeword length. According to a rough estimate, g values of 6, 8, and 10 are obtained from codeword lengths of 460, 4000, and 20,000, respectively. By increasing the codeword length, a possible large loop can be discovered in the parity-check matrix. However, longer codewords require increased circuit size. For 100 Gb/s class LSI, a codeword length g of 10 is very difficult to implement. Another way to increase g is to increase the redundancy. Regular LDPC codes having redundancies of 3, 10, and 35% enable their girths to be 6, 8, and 10, respectively. However, larger redundancy causes higher bit rate. In the case of 100 Gb/s, 35% redundancy cannot be allowed because its bit rate would reach as high as 140 Gb/s.

In slower-speed systems, e.g., mobile and wireless, one can use either of the above approaches, because their circuit implementation is not too difficult. However, in higher-speed optical communication systems, in particular at 100 Gb/s, very long codewords or large redundancy can never be allowed. Just as an example, a fully parallel implementation of an LDPC decoder for 10 G Ethernet is very large, requiring about 12,000 edge connections in each direction; if 6-bit representations of the probability values are used, then 72,000 bidirectional wires have to be routed between equality nodes and check nodes. To combat this hardware complexity problem, approximations and hardware reduction techniques have been developed.

6.5.4.3.2　Concatenated LDPC and RS codes

The preferred approach requires neither lengthening the codeword nor greatly increasing the redundancy to improve the union bound. It is possible to design LDPC codes without an error floor [58], but our solution is to apply an LDPC as an inner code, concatenating it with an outer RS code in order to reduce the residual errors left after LDPC decoding. An error burst of several tens of bits is sometimes observed following the LDPC decoding because of its advantage of circuit size rather than the design of LDPC codes only. Therefore, an RS code is purposely selected instead of a BCH code, because RS codes have higher tolerance to error bursts. We adopted a 2-bit soft decision for showing the effectiveness of the soft-decision decoding. Taking the restrictions of the OTU4V framing rules discussed in the ITU-T into account, an LDPC codeword length of 9216 codeword bits and 7936 information bits can be selected. RS (992, 956) was selected to minimize the overhead redundancy, this being adequate to suppress the undesired error floor. The total redundancy of the OTU4V-framed LDPC (9216, 7936) + RS (992, 956) codes is 20.5%. Figure 6.37 shows the experimental results measured for a prototype using FPGAs [65]. No error floor was seen thanks to the concatenated RS codes. The gross coding gain at

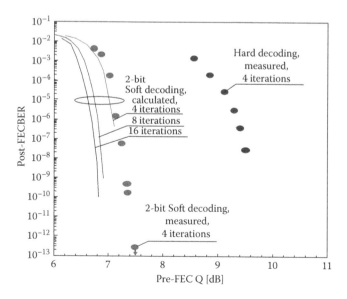

FIGURE 6.37　Experimental FEC performance results.

a post-FEC BER of 1×10^{-15} is 10.5 dB. Taking the increased bit rate into account, the NCG is 9.34 dB. The measured coding gain for hard decision was about 2.4 dB less than that obtained by the 2-bit soft decision. With 16 iterations and 2-bit soft decision, a further improved NCG of 10.3 dB can be expected at a post-FEC BER of 1×10^{-15}. With increasing 3 or more soft-decision bits, further improvement of error correction performance can be expected.

By using the new concept of triple-concatenated LDPC with block codes, further improvement of the error correction performance is possible. A computer simulation shows an NCG of 10.8 dB for DP-QPSK in a synchronously detected digital coherent system at 125 Gb/s [66].

6.5.4.3.3   Future Implementations

The relevant industry represented by OIF and ITU-T is trying to establish a common set of evaluation criteria in order to provide the fairest comparison of various different FEC codes useful for the consideration of commercialization. In addition to the critical gain parameter, i.e., NCG, other parameters, such as overhead rate, latency, encoding and decoding complexity, power consumption, implementation issues, etc., are also carefully considered. Soft-decision FEC needs not only an ultra-fast ADC, but also various follow-up complicated functions, such as a fast interface and huge parallelized circuit. So it's commonly believed that hard-decision FEC is the primary choice of FEC codes for the recommended 7% overhead that is currently ratified by ITU-T standards.

The beauty of hard-decision FEC devices is the ease to operate directly with such discrete components as EDC, CDR, and SerDes devices, without the need of fast speed interface required by soft-decision FEC. Early Vitesse studies on 40 Gb/s transmission, conducted prior to CI-BCH implementation, demonstrated the feasibility of combining eFEC/EDC design with the kind of SerDes devices developed by Sierra Monolithics, Inc. in the form of a 40 G 300-pin module [45]. When Sierra used IBM's 7 HP SiGe BiCMOS process, the SerDes was implemented in a hybrid package; migration to 8 HP allowed full DQPSK multiplexing and demultiplexing to be implemented in a surface-mount package with a 40% reduction in power dissipation.

In preparation for standard implementations of CI-BCH, Vitesse designed a 2.5 Gb/s test platform in a single FPGA, as described above. Studies conducted indicate that a 40 Gb/s encoder/decoder with 256-bit bus width could be implemented in roughly one-half of an LXT330 FPGA. Vitesse subsequently combined an encoder, decoder, and random error generator within a single FPGA supporting 43 Gb/s OTU3 operation. Simulated studies of CI-BCH coding implemented in different families of FPGA have demonstrated that a 100 Gb/s encoder/decoder, incorporating 200,000 flip-flops and 330,000 logic units, could be implemented in a similar LXT330 class device, using optimal design tools.

For higher overhead, soft-decision FEC will play a role similar to a higher-overhead hard-decision FEC whenever higher NCG is required [58, 66]. Soft-decision FEC has the advantage to provide the extra gain over hard-decision FEC at the expense of encoder and decoder complexity. Because of the fast interface needed to connect the DSP, the soft-decision FEC is normally bonded into the module, and forced to be implemented in the same LSI as DSP. Mitsubishi's work with an

FPGA-based emulator indicates a 100 Gb/s soft-decision FEC, with NCG in excess of 10 dB, could be implemented in 160 million gates. Given 45 nm CMOS process trends, a single-chip soft-decision FEC appears feasible in the near future.

6.6 CONVERGENCE OF PHOTONICS WITH ELECTRONICS

As complexity can often lead to added cost through a higher number of components and manufacturing yield issues, an important design constraint is how to meet the market cost targets. To achieve the cost targets, the use of electronic integration and photonic integration can be implemented to reduce component count and improve manufacturability. The integration of OTU4 OH processing, SD-FEC, DSP, firmware, and control loops really creates a 100 Gb/s line-side "system in a module" concept, where other external components such as ASICs, optical compensators, optical amplifiers, etc., are not required on the transponder host board.

6.6.1 ELECTRONIC INTEGRATION

The monolithic integration of ADC, DSP, and FEC in a single CMOS chip provides good cost reduction from high functionality concentrated into a single chip. It also removes the need for >Tb/s bus speed interconnect between ADC > DSP and DSP > FEC chips, which saves in space and power. By coding 4 bits/symbol, this reduces the baud rate to a level that can be implemented in CMOS technology. This use of CMOS EDC rather than optical CD/PMD dispersion compensators saves considerable cost in reduced number of components and also in space and power. The use of SD-FEC, approaching Shannon's limit in terms of OSNR sensitivity, increases the optical reach, which in turn reduces (or eliminates) the need for costly OEO regeneration and improves network economics.

6.6.2 PHOTONIC INTEGRATION

The OIF is promoting standardization of the photonic integration for both the 100 Gb/s transmit and receive optical subassemblies. On the photonics transmit side, a single optical assembly contains the nested MZMs, PBS, and splitters. On the photonics receive side, a single optical assembly houses the PBS, phase hybrids, balanced photodetectors, and linear transimpedance amplifiers (TIAs). This OIF standardization has helped to create a supplier ecosystem around this technology, standardizing modulation format, module MSA footprint, electrical and control interfaces, and Tx and Rx integrated photonics blocks. This concentrates R&D investment, brings a cost reduction through photonic integration, and enables multisourcing at the integrated photonics and module level. These all drive economic gain for the system vendors and service providers.

Integration of these optical functions is viewed as a key requirement for ensuring market viability with promise to further converge with electronics integration [67]. This view is shared by many in the industry, as evidenced by the large support of efforts in industry forums such as the OIF, where component and module suppliers are involved in defining key aspects of practical implementations.

6.7 SUMMARY AND OUTLOOK

This chapter overviews the development status and enabling technologies of the 100 G transceivers and transponders, especially the 100 G form factor pluggable (CFP) and the OIF 168-pin transponder MSA for 100 GbE client optics interconnection and transport applications. There exists tremendous ongoing effort to design compact and cost-effective modules toward cheaper 100 G optical ports, such as CFP2/4 and QSFP28 for either direct or coherent detection.

The continual worldwide growth in smart phone and tablet data traffic as well as other broadband mobile devices is driving global network capacity expansion. The optical channel capacity has had to grow to keep up with this traffic demand over the past several decades. With commercial 100 G platforms being deployed today, the need for >100 G becomes obvious and the industry starts a debate on 400 G vs. 1 T for the next date rate. It's estimated that the bandwidth demand will reach a maximum of 400 G in 2015 and will approximately reach 1 T in 2020.

IEEE has launched the new IEEE 802.3 400 Gbps Ethernet Study Group to explore the development of a 400 G Ethernet standard. Likewise, the OIF Physical and Link Layer (PLL) Working Group has launched a new project to define a module interface implementation agreement (IA) for 400 G long-haul optical transmission. Figure 6.38 shows the timeline for development efforts of next-generation standardized Ethernet speeds, with an expected timeline for 400 G Ethernet with standard ratification around 2017 [68]. Both those efforts, with primary focus on the 400 G physical layer specifications, will provide the industry with technology parameters of choice for near-term component development and implementation strategies.

It's well known that the present optical transport data rates are limited by IC electronics speed. From 100 G design experience, a higher data rate is achievable by coding multiple bits per symbol and using coherent detection and m-ary quadrature amplitude modulation (M-QAM) schemes. Multiple modulation formats would be

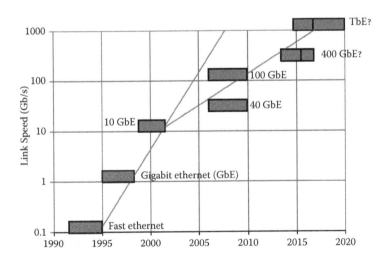

FIGURE 6.38 The next-gen standarizing Ethernet speeds.

desirable at 400 Gbit/s, to allow for trade-offs between spectral efficiency and reach. One option would combine four channels of 100 Gbit/s apiece into one 400 Gbit/s superchannel using DP-QPSK modulation. This would have relatively long reach but would take up 150 GHz of spectrum. The other option would be two 200 Gbit/s channels using 16-QAM modulation, taking up only 75 GHz but sacrificing some distance, due to the constellation OSNR penalty.

Besides the telecom applications, there will also be a need for much shorter distance at 400 G in datacom systems. Current 100 G products using 10:4 gearbox SerDes with individual lanes running at 25 Gbps are just hitting the market; the costs of bundling more than 16 lanes is likely prohibitive. Keeping in mind the cost per bit must continue to fall dramatically, developing 400 G at the right cost tends to consider the options using modulation schemes, including PAM and new forms of enhanced FECs. Various different Ethernet distances are required inside the data centers, which are combined with the considerations to support optical transport networks, Energy Efficient Ethernet, and auto-negotation of data rates while keeping latency low.

The new 100 G and upcoming 400 G transceiver designs bring up significant increases in transmitter and receiver complexity and create many different ways of building optics for the fabrication of lasers, waveguides, mux/demux, splitting/combiners, gratings/filters, detectors and modulators, etc. In addition to the electronics integration, photonic integration will play a critical role as it does the same for the 10 G design. As a disruption technology, silicon photonics based on photonics and electronics convergence enters into the stage of engineering innovations and product development, holding promise for the future with lower cost, size, and power consumption, but it has its own technical and market hurdles to overcome while competing on performance against existing established technologies.

REFERENCES

1. Moore's law, http://www.mooreslaw.org/.
2. R. Seifeit, *The Switch book: The complete guide to LAN switching technology*, John Wiley & Sons, New York, 2000.
3. IEEE 802.3 HSSG Tutorial 1107, November 2007, http://www.ieee802.org/3/hssg/public/nov07/HSSG Tutorial 1107.zip.
4. IEEE 802.3ba 40 Gb/s and 100 Gb/s Task Force (formerly Higher Speed Studay Group (HSSG)), http://ieee802.org/3/hssg/public/index.html.
5. ITU-T SG15, http://www.itu.int/ITU-T/studygroups/com15/index.asp.
6. O. Ishida, 40/100GbE technologies and related activities of IEEE standardization, in *OFC/NFOEC 2009*, Paper OWR5.
7. Optical Internetworking Forum (OIF), www.oiforum.com.
8. Z. Hatab and F. Chang, 100G electrical to optical interfaces, ETS 2010 invited talk.
9. OIF SFI4-1 IA, OIF-SFI4-01.0, http://www.oiforum.com/public/documents/OIF-SFI4-01.0.pdf.
10. R. M. Metcalfe et al., U.S. patent 4,063,220, Ethernet patent, Multipoint data communication system with collision detection, 1977.
11. Digital Equipment Corporation, Intel Corporation, and Xerox Corporation, *The Ethernet, a local area network. Data link layer and physical layer specifications*, version 1.0, Xerox Corporation, September 30, 1980, http://ethernethistory.typepad.com/papers/EthernetSpec.pdf (accessed December 10, 2011).

12. Digital Equipment Corporation, Intel Corporation, and Xerox Corporation, *The Ethernet, a local area network. Data link layer and physical layer specifications*, version 2.0, Xerox Corporation, November 1982, http://decnet.ipv7.net/docs/dundas/aa-k759b-tk.pdf (accessed December 10, 2011).

13. P. Winzer, Beyond 100G Ethernet, *IEEE Commun. Mag.*, 48(7), 26–30, 2010.

14. J. Anderson, Optical transceivers for 100GbE and its transport, *IEEE Commun. Mag.*, 48(3), S35–S40, 2010.

15. Extreme Networks, *How the port density of a data center LAN switch impacts scalability and total cost of ownership*, white paper, June 4, 2012, http://www.extremenetworks.com/libraries/whitepapers/How_Port_Density_of_a_DC_LAN_Switch_Impacts_Scalability_and TCO.pdf.

16. C. Cole, Next generation CFP modules, in *OFC 2012*, Paper NTu1F.1.

17. CFP MSA, CFP MSA hardware specifications, http://www.cfp-msa.org/.

18. IEEE 802.3 Bandwidth Ad Hoc, http://www.ieee802.org/3/ad_hoc/bwa/.

19. A. Bach, High speed networking and the race to zero, presented at Hot Interconnects (HOTI) Conference, New York, NY, August 25–27, 2009, http://www. hoti.org/hoti17/program/.

20. A. Bach, The financial industry's race to zero latency and terabit networking, presented at OFC 2010, keynote speech for service provider summit.

21. C. E. Shannon, A mathematical theory of communication, *Bell System Tech. J.*, 27, 379–423, 623–656, 1948.

22. G. Wellbrock and T. J. Xia, The road to 100g deployment (commentary), *IEEE Commun. Mag.*, 48(3), S14–S18, 2010.

23. F. Chang et al., Forward error correction for 100G transport networks, *IEEE Commun. Mag.*, 48(3), S48–S55, 2010.

24. S. Gringeri, B. Basch, T. J. Xia, Technical considerations for supporting data rates beyond 100 Gb/s, *IEEE Commun. Mag.*, 50(2), S21–S30, 2012.

25. F. Chang, Consider IC challenges for 40/100G system design reality check: views from the physical layer, presented at 19th Annual Wireless and Optical Communications Conference (WOCC), May 14–15, 2010.

26. F. Chang and S. Shang, High speed SerDes and EDC/FEC for 40G/100G optical communications, presented at Asia Communications and Photonics Conference and Exhibition (ACP), ACP 2009, 40G/100G workshop "Meeting the Need for Speed: 40G and the Road to 100G," Shanghai, China, September 2009.

27. Y. Ito, A new paradiagram in optical communication and networks, *IEEE commun Mag.*, 51(3), 24–26, 2013.

28. H. Toyoda et al., 100GbE PHY and MAC layer implementation, *IEEE Commun. Mag.*, 48(3), S41–S47, 2010.

29. X. Zhang et al., The application scheme of 40G and 100G Ethernet, presented at 10th International Conference on Optical Communications and Networks (ICOCN 2011), November 5, 2011–December 7, 2011.

30. J. Roese et al., Optical transport networks evolving with 100 Gigabit Ethernet, *IEEE Commun. Mag.*, 48(3), S28–S34, 2010.

31. T. Ohara and O Ishida, Standardization activities for the optical transport networks, *NTT Tech. Rev.*, 7(3), 1–6, 2009.

32. P. Winterling, *100 Gigabit Ethernet—fundamentals, trends and measurement requirements*, JDSU white paper, http://www.jdsu.com/ProductLiterature/100GE-fundamentals-white paper.pdf.

33. CFP MSA, CFP MSA hardware specifications, http://www.cfp-msa.org/.

34. F. Chang et al., Forward error correction for 100G transport networks, *IEEE Commun. Mag.*, 48(3), S48–S55, 2010.

35. J. Hong et al., 40G and 100G modules enable next-generation networks, *Proc. SPIE*, 763115, 1–7, 2009.

36. Z. Hatab and F. Chang, Technical feasibility of serial 25 Gb/s signaling over backplanes: channel and system analysis, presented at IEEE 802.3 100GCU Study Group, January 12–13, 2011, http://www.ieee802.org/3/100GCU/public/jan11/hatab 01 0111.pdf.
37. *CEI-28G: paving the way for 100 gigabit*, OIF Physical and Link Layer (PLL) white paper, 2009.
38. S. Hagene, Tyco gen 2 VSR channel models, OIF Physical and Link Layer (PLL) Working Group presentation, July 2010.
39. S. Hagene, Gen 2 VSR channel test results, OIF Physical and Link Layer (PLL) Working Group presentation, July 2010.
40. W. T. Beyene, The design of continuous-time linear equalizers using model order reduction techniques, in *Proceedings of the IEEE 17th Topical Meeting on Electrical Performance of Electronic Packaging (EPEP)*, San Jose, CA, October 21–24, 2008, pp. 187–190.
41. Z. Hatab and F. Chang, CEI-28G-VSR channel simulations and system SNR, OIF Physical and Link Layer (PLL) Working Group presentation, July 2010.
42. S. Haykin, *Adaptive filter theory*, Upper Saddle River, NJ, Prentice Hall, 2001, chap. 9.
43. M. Harwood et al., A 225mW 28Gb/s SerDes in 40nm CMOS with 13dB of analog equalization for 100GBASE-LR4 and optical transport lane 4.4 applications, in *ISSCC 2012*, Paper 19.2.
44. M. Kono et al., A 400-Gb/s and low-power physical-layer architecture for next-generation Ethernet, presented at IEEE ICC 2011.
45. H. Toyoda et al., 100GbE PHY and MAC layer implementation, *IEEE Commun. Mag.*, 48(3), S41–S47, 2010.
46. F. Chang, Application aspects of enhanced HD-FEC for 40/100G systems, ECOC workshop WSII, Torino, Italy, September 19, 2010.
47. P. Bower and I. Dedic, High speed converters and DSP for 100G and beyond, *Opt. Fiber Technol.*, 2011, doi: 10.1016.
48. Y. Cai, Performance limits of FEC and modulation formats in optical fiber communications, in *IEEE-LEOS 2006*, October 2006, Paper WH1.
49. T. Mizuochi, Soft-decision FEC for 100 Gb/s DSP based transmission, in *IEEE Photonics Society Summer Topical Meetings Digest*, Newport Beach, CA, July 2009, Paper WC1.1, pp. 107–108.
50. W. D. Grover, Forward error correction in dispersion limited lightwave systems, *J. Lightwave Tech.*, 6(5), 643–645, 1988.
51. S. Yamamoto et al., 5 Gb/s optical transmission terminal equipment using forward error correcting code and optical amplifier, *IEE Electronics Lett.*, 30(3), 254–255, 1994.
52. G. D. Forney, *Concatenated codes*, MIT Press, Cambridge, MA, 1966.
53. Y. Kisaka et al., Fully transparent multiplexing and transport of 10GbE-LANPHY signals in 44.6-Gb/s-based RZ-DQPSK WDM transmission, in *OFC/NFOEC 2007*, Anaheim, CA, March 2007, Paper OThL1.
54. A. Puc et al., Concatenated FEC experiment over 5000 km long straight line WDM test bed, in *OFC 1999*, San Diego, CA, February 1999, Paper ThQ6-1, pp. 255–258.
55. T. Mizuochi et al., Experimental demonstration of net coding gain of 10.1 dB using 12.4 Gb/s block turbo code with 3-bit soft decision, in *OFC 2003*, Atlanta, GA, March 2003, Paper PD21.
56. K. Ouchi et al., A fully integrated block turbo code FEC for 10 Gb/s optical communication systems, in *OFC/NFOEC 2006*, Anaheim, CA, March 2006, Paper OTuK4.
57. R. G. Gallager, Low-density parity-check codes, *IRE Trans. Info. Theory*, IT-8(1), 21–28, 1962.
58. I. B. Djordjevic et al., Next generation FEC for high capacity communication in optical transport network, *IEEE J. Lightwave Tech.*, 27(16), 3518–3530, 2009.
59. M. Scholten et al., Proposed OTU4 FEC, presented at ITU-T Q11/SG15 interim meeting, March 2009.

60. T. J. Richardson and R. L. Urbanke, The capacity of low-density parity-check codes under message-passing decoding, *IEEE Trans. Info. Theory*, 47(2), 599–618, 2001.
61. M. Scholten et al., Continuously-interleaved BCH (CIBCH) FEC delivers best in class NECG for 40G and 100G metro applications, in *OFC/NFOEC 2010*, San Diego, CA, March 2010, Paper NTuB3.
62. D. J. C. MacKay, Good error correcting codes based on very sparse matrices, *IEEE Trans. Info. Theory*, 45(2), 399–431, 1999.
63. Z. Zhang et al., Lowering LDPC error floors by postprocessing, presented at IEEE GLOBECOM, New Orleans, LA, November 2008.
64. P. Lee et al., Error floors in LDPC codes: fast simulation, bounds and hardware emulation, in *IEEEISIT 2008*, July 2008, pp. 444–448.
65. T. Mizuochi et al., Experimental demonstration of concatenated LDPC and RS codes by FPGAs emulation, *IEEE Photonics Tech. Lett.*, 21(18), 1302–1304, 2009.
66. Y. Miyata et al., A triple-concatenated FEC using soft-decision decoding for 100 Gb/s optical transmission, in *OFC/NFOEC 2010*, San Diego, CA, March 2010, Paper OThL3.
67. P. D. Dobbelaere et al., Si photonics based high-speed transceivers, in *ECOC 2012*, Paper We.1.E.5.
68. IEEE 802.3 400 Gb/s Ethernet Study Group, http://www.ieee802.org/3/400GSG/public/.

7 Equalization and Multilevel Modulation for Multi-Gbps Chip-to-Chip Links

Anthony Chan Carusone

CONTENTS

7.1 INTRODUCTION

To avoid becoming a bottleneck in large electronic systems, the bit rates that must be communicated over chip-to-chip communication links are increasing exponentially. Traditionally, the data are communicated over printed circuit board (PCB) traces that have accommodated increasing bit rates with simple circuits and advancing process technologies. However, links operating at speeds exceeding 1 GHz begin to encounter the inherent bandwidth limitations of the interconnect.

One approach to overcoming the bandwidth limitations of chip-to-chip links is to use alternative interconnect technologies. This may include the use of low-loss materials for PCB fabrication, optical [1], or even wireless [2] interconnect technologies. However, the cheapest method for increasing bit rates over existing chip-to-chip links remains low-power electronic signal processing integrated into transceivers at either end, whenever this is possible.

Well-known techniques have been used for this purpose, including transmit equalization [3–5], receive-side equalization [6–8], decision feedback equalization [4, 9, 10], and multilevel modulation [3, 4, 8]. They are used for both serial [3, 8, 11] and parallel [9, 12, 13] chip-to-chip links at bit rates from 1 to 10 Gb/s. Crosstalk is generally identified as a major source of interference in these links [8, 12, 14]. Crosstalk cancellation is possible only when the aggressors are on-chip [8, 12], often at the expense of increased power, complexity, and decreased signal swing when low supply voltages offer little headroom.

This chapter will identify major impairments to multi-Gbps chip-to-chip links, describe techniques to model them, and survey the methods being used to mitigate those impairments and enable robust links.

First, typical chip-to-chip links are described and categorized in Section 7.2. Then, models applicable to multi-Gbps chip-to-chip links are developed in Section 7.3. In Section 7.4, equalization is described, including its implementation at the transmitter and receiver. Section 7.5 compares different modulation schemes. A common criterion for selecting an appropriate modulation scheme is examined and found to be oversimplified. Finally, a summary and conclusions are presented in Section 7.6.

7.2 APPLICATIONS

In this chapter, we are concerned with links carrying traffic at data rates from 1 Gbps to 100 Gbps (measured per channel in the case of a parallel bus). At these rates, an interconnect of even just a few centimeters in length has a propagation delay that is appreciable compared to a bit period. Hence, PCB traces are designed to have a specific characteristic impedance, and termination is provided at both transmitter and receiver.

State-of-the-art fully differential chip-to-chip links remain more than one order of magnitude faster than their single-ended counterparts. This more than offsets the 2× increase in the number of pins and traces required to accommodate them. Hence, this chapter will focus on fully differential links, which are used almost invariably for high-performance chip-to-chip applications.

The most popular PCB trace geometries for implementing controlled-impedance chip-to-chip transmission lines are the microstrip and stripline, both shown

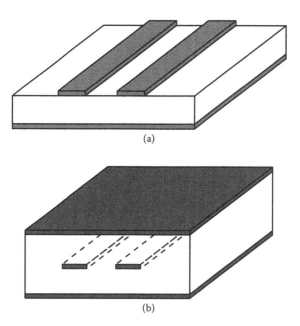

FIGURE 7.1 Popular PCB trace geometries for high-performance chip-to-chip communication: (a) a microstrip transmission line and (b) a stripline transmission line.

in Figure 7.1. The microstrip may be lower cost than the stripline since it occupies only two layers on the surface of the PCB. However, the electromagnetic fields in the stripline are completely confined to a uniform dielectric medium, making it less sensitive to environmental variations than the microstrip whose electrical characteristics may, for example, vary depending on the coating applied to the surface of the PCB.

The vast majority of high-performance chip-to-chip links may be broadly categorized into two groups:

- *Parallel chip-to-chip buses*: Several channels in parallel provide a high-throughput communication link between two integrated circuits (ICs) on the same PCB.
- *Serial links over backplanes*: These links comprise only a single electrical connection over which all data traffic is aggregated. They are generally longer and traverse more discontinuities than parallel buses such as connectors.

The nature and relative severity of the various channel impairments differ significantly between the two groups.

7.2.1 PARALLEL BUSES

A typical parallel chip-to-chip bus is depicted in Figure 7.2. It is typically characterized by multiple interconnects densely routed in parallel between integrated circuits and punctuated by relatively few discontinuities such as connectors. Such links are required, for example, between a central processing unit and a peripheral bridge or

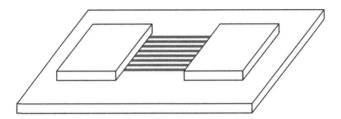

FIGURE 7.2   A parallel chip-to-chip bus.

co-processor, between processor and memory, or between high-speed data convert-ers and digital signal processors. The number of parallel connections varies widely depending on the application, of course, from 4 to 128 or more.

Transceivers for these applications must consume low power and low area. All of the parallel data streams are synchronized to the same clock, thereby permitting some of the power or area overhead associated with clocking to be amortized over the width of the bus. The most stringent channel impairments are crosstalk from neighboring transceivers, and in the case of very high-speed links, intersymbol interference (ISI).

At present, consumer products routinely include parallel bus links at data rates exceeding 1 Gbps, and data rates [15] exceeding 20 Gbps per transceiver pair have been demonstrated in experimental platforms [16].

7.2.2 SERIAL LINKS OVER BACKPLANES

Several important applications require a high-speed link between ICs on separate PCBs, for example, the links connecting high-speed physical layer transceivers and routers inside Internet switches, or between daughtercards in a high-performance parallel-processing computer. Wide parallel buses connecting PCBs require large connectors that are expensive and physically large. An obvious alternative is to aggregate the traffic from the entire bus onto a single serial link. An example of such a link is shown in Figure 7.3, where transmitter and receiver are separated by two connectors and three lengths of PCB trace. The "backplane" generally refers to the intermediate PCB carrying traffic between the two daughtercards, but similar design challenges arise in applications where transmitter and receiver are separated by only one connector and two lengths of PCB trace.

Another common requirement in serial links is to provide electrical isolation between transmitter and receiver. A series-connected AC-coupling capacitor is gener-ally provided for this purpose. AC-coupling allows the transmitter and receiver to oper-ate with different common-mode voltages, easing interoperability and providing an extra degree of freedom in the design of transceivers. However, the discontinuity pre-sented by the coupling capacitor, including the vias that may be required to access the surface of the PCB, represents yet another discontinuity in the link. It also introduces baseline wander in the presence of long strings of consecutive identical bits, and is therefore usually accompanied by a DC-balanced line code such as 8b10b coding [17].

Clearly, in comparison with the short parallel buses described above, backplane serial links face significantly tougher bandwidth limitations due to the discontinuities.

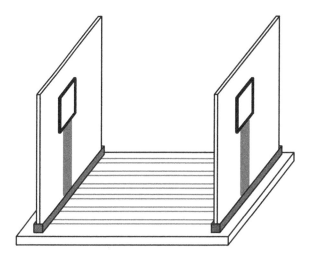

FIGURE 7.3   A serial link over a backplane.

7.3 LINK MODELING

This section briefly reviews the modeling of point-to-point chip-to-chip links. A complex chip-to-chip link, such as the serial link depicted in Figure 7.3, may comprise several sections of stripline or microstrip transmission line with mismatched terminations and IC packaging at either end and punctuated by vias and connectors along the way. The frequency response of a link is very sensitive to the discontinuities, so they must be included in link models.

The PCB traces and discontinuities are linear two-port networks. The approach taken in this section is to obtain a linear two-port description of each transmission line segment and discontinuity, and then cascade them to generate a model of the entire link. Unfortunately, the two-port descriptions of various portions of the link are often specified in different formats. Hence, all two-port descriptions must be converted into a common form before combining them to construct the complete link model. Transmission matrices are a particularly useful two-port description since they are easily cascaded.

Section 7.3.1 provides some background on electrical two-port networks and transmission matrices. Section 7.3.2 then provides a scalable two-port description of PCB interconnect, including skin effect and dielectric losses. Section 7.3.3 follows with examples of complete links modeled in this way.

7.3.1 LINEAR TWO-PORT NETWORKS

A general linear two-port electrical network is shown in Figure 7.4. The quantities $V_1(j\omega)$, $I_1(j\omega)$, $V_2(j\omega)$, and $I_2(j\omega)$ are complex-valued phasor representations of the voltages and currents at ports 1 and 2 of the network, respectively, at angular frequency $\omega = 2\pi f$. A transmission matrix $\Phi(j\omega)$ is 2×2 with frequency-dependent

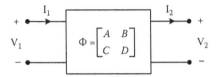

FIGURE 7.4 A general linear two-port electrical network and its transmission matrix.

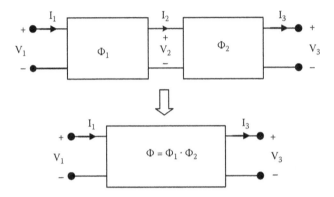

FIGURE 7.5 A series connection of linear two-port networks.

complex-valued entries $A(j\omega)$, $B(j\omega)$, $C(j\omega)$, and $D(j\omega)$ that relate the port voltages and currents.

$$\begin{bmatrix} V_1(j\omega) \\ I_1(j\omega) \end{bmatrix} = \begin{bmatrix} A(j\omega) & B(j\omega) \\ C(j\omega) & D(j\omega) \end{bmatrix} \begin{bmatrix} V_2(j\omega) \\ I_2(j\omega) \end{bmatrix} = \Phi\,(j\omega) \begin{bmatrix} V_2(j\omega) \\ I_2(j\omega) \end{bmatrix} \qquad (7.1)$$

The matrix Φ provides a complete description of a linear electrical two-port network. Electrical two-ports may otherwise be described by

- Impedance matrices
- Admittance matrices
- Scattering parameters

Any of these may be straightforwardly transformed into transmission matrices via the expressions provided in the appendix.

The transmission matrix of series-connected two-ports is simply the product of the transmission matrices of the constituent two-ports. Figure 7.5 shows a simple example of two two-port networks connected in series having transmission matrices $\Phi_1(j\omega)$ and $\Phi_2(j\omega)$, respectively. The voltage and current phasors at the far ends of the series connection may be written in terms of $\Phi_1(j\omega)$ and $\Phi_2(j\omega)$, resulting in a transmission matrix describing the entire link, $\Phi(j\omega)$,

$$\begin{bmatrix} V_1(j\omega) \\ I_1(j\omega) \end{bmatrix} = \Phi_1(j\omega) \begin{bmatrix} V_2(j\omega) \\ I_2(j\omega) \end{bmatrix} = \Phi_1(j\omega)\Phi_2(j\omega) \begin{bmatrix} V_3(j\omega) \\ I_3(j\omega) \end{bmatrix} \equiv \Phi \begin{bmatrix} V_3(j\omega) \\ I_3(j\omega) \end{bmatrix} \quad (7.2)$$

This result is straightforwardly generalized, so that the transmission matrix of any number of two-ports connected in series is given by the product of the constituent two-ports,

$$\Phi(j\omega) = \prod_k \Phi_k(j\omega) \quad (7.3)$$

A transmission matrix for the entire link may be built in this way from transmission matrices for the terminations at either end, discontinuities such as vias and connectors, and the intervening sections of lossy PCB trace. This will be done in Section 7.3.3.

7.3.2   THE TRANSMISSION LINE

Whereas the discontinuities of a link are complex structures best characterized using either electromagnetic simulation or direct measurement, PCB traces for high-performance chip-to-chip links employ transmission line geometries for which scalable and accurate models are available. This section describes such a scalable model and develops the corresponding transmission matrices. The advantages of such scalable models (as opposed to direct measurement) are manifold. Parameters such as the length of the link, the location of discontinuities, and the loss tangent can be easily swept to evaluate the performance and robustness of a given transceiver in the complete variety of channels likely to be encountered in the field. Furthermore, they provide an exact result with no measurement noise. The best approach is to use measurement results from a few test cases to refine a scalable model of the link.

For high-performance chip-to-chip communication, wired interconnects are transmission lines that may be modeled using a quasi-static transverse electromagnetic (TEM) approximation. In general, the resulting model comprises frequency-dependent values of resistance, inductance, conductance, and capacitance per unit length, as shown in Figure 7.6.

The frequency dependence of R, L, G, and C can be described in terms of a few parameters to model skin effect and dielectric losses [18].

The skin effect is modeled by a complex-valued frequency dependence in R that includes the conductors' DC loss per unit length (R_0, often negligible for short traces) and a skin effect constant (R_S),

$$R(f) = R_0 + \sqrt{f} \cdot (1 + j)R_S \quad (7.4)$$

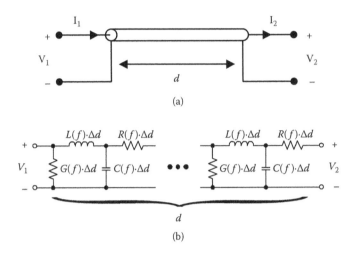

FIGURE 7.6   (a) A general transmission line of length d and (b) a quasi-static TEM approximate model.

In combination with a frequency-independent value for series inductance (L_0), this results in a loss that increases at 10 dB/decade.

Conduction through the dielectric is generally negligible, $G_0 = 0$. But, alternating electromagnetic fields cause heating and, hence, loss in the dielectric that increases with frequency. This is modeled using a complex-valued frequency-dependant per-unit-length capacitance,

$$C(f) = C_0 \cdot \left(\frac{jf}{f_0} \right)^{\frac{-2\theta}{\pi}} \tag{7.5}$$

Here, C_0 is the transmission line's capacitance per unit length at low frequencies and θ and f_0 are parameters that provide for linearly increasing (20 dB/decade) losses at very high frequencies.

Since dielectric losses impose strict bandwidth limitations on chip-to-chip communication, there is strong motivation to consider low-loss materials such as Teflon for this purpose. However, the epoxy laminate FR4 remains a popular dielectric material in spite of its relatively high loss due to its mechanical robustness and low cost. Once the frequency-dependent values of R, L, G, and C are known, they can be converted to transmission matrices using formulas in the appendix.

7.3.3   Complete Links

The transmission matrix of PCB traces can be combined with those of discontinuities using Equation (7.3) to make a complete link model. Matlab functions are useful for combining two-port descriptions of various linear networks into a single transmission matrix, and then performing the required (IFFT) to obtain a time-domain representation, are readily available Matlab functions and were used for the examples in this section.

The interconnect model is a differential microstrip with 100 Ω matched terminations on an FR4 dielectric. Model parameter values can be obtained from analytical expressions for a known trace geometry, electromagnetic simulations, or fitting to real measured results. The values used here are typical of a 100 Ω differential impedance microstrip on an FR4 PCB and are given in Table 7.1.

Figure 7.7 shows the influence of a simple package model on the frequency and transient responses of a link between two ICs on the same PCB. A 5 cm link has a 3 dB bandwidth exceeding 20 GHz without including IC packaging, but the 3 dB bandwidth drops to just over 2.5 GHz with the inclusion of a simple package model. The reflection due to the package's discontinuity is clearly visible in the step response, and there is a large ripple in the frequency response.

Over longer channels, the effects of PCB trace losses become more significant than the effects of discontinuities. For example, introducing package models into the 20 cm link reduces the 3 dB bandwidth by less than 25%, from 2.2 GHz to 1.7 GHz. This is sensible since the reflections introduced by the packaging must traverse a longer length of PCB trace before appearing at the receiver and, hence, are attenuated by the trace losses. Hardly any reflections are visible in the 50 cm transient response. However, the skin effect and dielectric losses are large, contributing almost 25 dB loss at 10 GHz.

In Figure 7.7, the only discontinuities are the IC packages at either end of the link. This is the case in, for instance, point-to-point parallel links between two ICs on the same PCB. However, serial links over a backplane have more discontinuities, most notably the connectors between the motherboard and daughtercards. Measured responses for two real backplanes are plotted in Figure 7.8. Both channels have the same daughtercards with 10 cm of interconnect each and the same backplane connectors; one channel has a 10 cm length of interconnect on the motherboard, and the other has 40 cm of interconnect on the motherboard. Even though they are very different lengths, both channels have similar responses because the loss and reflections due to connectors are the dominant effect.

TABLE 7.1

Model Parameter Values Typical of a 100 Ω Differential Impedance Interconnect on an FR4 PCB

R_0 (Ω/m)	≈ 0
R_S (Ω / m · $\sqrt{\text{Hz}}$)	$8.7 \cdot 10^{-9}$
L_0 (nH/m)	817
G_0 (Ω^{-1}/m)	0
C_0 (pF/m)	148
θ_0	0.022
f_0 (GHz)	10

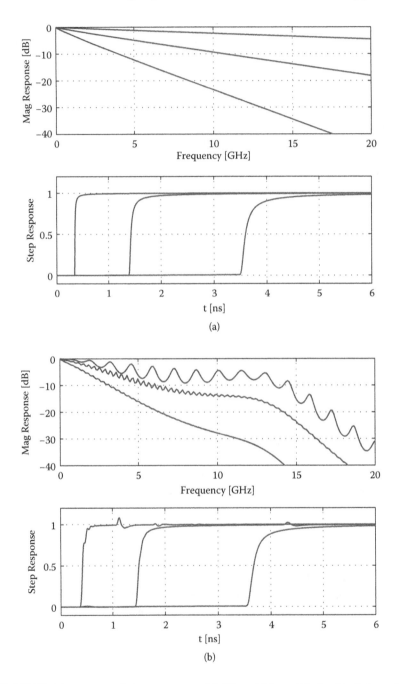

FIGURE 7.7 Frequency and step responses of 5, 20, and 50 cm chip-to-chip links on FR4 dielectric (a) without and (b) with simple IC package models included at either end.

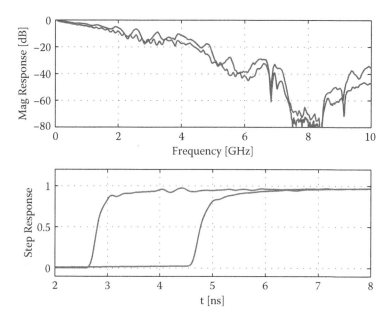

FIGURE 7.8   Measured frequency and step responses of two real backplane links comprising 10 and 40 cm motherboards and two 10 cm daughtercards.

7.4   EQUALIZATION

It is well established that equalization is necessary for multi-Gb/s communication over many practical chip-to-chip links. A wide variety of filters have been employed at both the transmitter and the receiver for this purpose.

Figure 7.9 presents a system model for the linear portion of a chip-to-chip link. It comprises a transmit filter with frequency response $B(\omega)$ and impulse response $b(t)$ operating on a transmit signal $u(t)$, a channel with frequency response $H(\omega)$ and impulse response $h(t)$, and a receiver with frequency response $C(\omega)$ and impulse response $c(t)$. Equalization may be incorporated into the transmitter $B(\omega)$ or the receiver $C(\omega)$.

Recently, both transmitter and receiver filters have been combined with decision feedback to equalize long or challenging links [19–21]. With such a wide variety of equalizer architectures in use, there is great interest in comparing their performance to arrive at the optimal design for a given link.

7.4.1   TRANSMIT EQUALIZATION

Practical considerations often limit the peak output voltage of the transmitter to a constant L. Hence, the transmit filter must satisfy the following constraint for all possible inputs $u(t)$:

$$|(u*b)(t)| = \left| \int u(t-t')b(t') \cdot dt' \right| \leq L \tag{7.6}$$

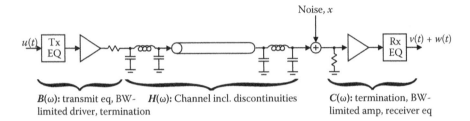

$B(\omega)$: transmit eq, BW- $H(\omega)$: Channel incl. discontinuities $C(\omega)$: termination, BW-
limited driver, termination   limited amp, receiver eq

FIGURE 7.9   System model of the linear portions of a chip-to-chip link.

It is always possible to choose an input $u(t)$ for which $|\int u(t-t')b(t')\cdot dt'| = \int |u(t-t')|$ $|b(t')|\cdot dt'$ So, to ensure (7.6) is satisfied for any input $u(t)$, the following expression must also hold:

$$\int |u(t-t')||b(t')|\, dt' \leq L \tag{7.7}$$

Without loss of generality, we may assume that the input to the transmit filter is normalized to have a signal swing within the range $|u(t)| \leq 1$. Therefore, the transmit filter must satisfy the following equation:

$$\int |b(t')|\, dt' \leq L \tag{7.8}$$

Multiplying by $|e^{j\omega t'}| = 1$,

$$\int |b(t')|\cdot dt' = \int |b(t')|\left|e^{j\omega t'}\right|\cdot dt' \tag{7.9}$$

$$\geq \left|\int b(t')e^{j\omega t'}\cdot dt'\right| = |B(\omega)| \tag{7.10}$$

Combining Equations (7.8) and (7.10) gives

$$|B(\omega)| \leq L \tag{7.11}$$

Equation (7.11) must hold for all ω if the transmit waveform's peaks are to be limited in amplitude to the constant L. The simplest transmitter satisfying (7.11) is an amplifier with constant gain L. If any spectral shaping is to be introduced at the transmitter to compensate for frequency-dependent channel losses, it is done by reducing the gain at some frequencies below L, in accordance with (7.11). Hence, peak-constrained transmit equalization is often referred to as de-emphasis since it equalizes the channel response by attenuating the low-frequency portion of the transmit signal spectrum below the level attainable with a simple transmitter having a constant gain L.

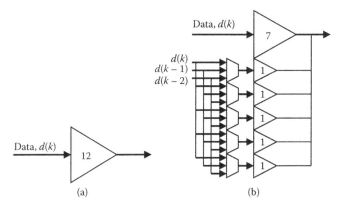

FIGURE 7.10 (a) A conventional transmitter without de-emphasis. (b) A modification to provide de-emphasis by partitioning the output driver into 12 segments.

The filter $B(\omega)$ in many chip-to-chip links is a finite impulse response (FIR) filter. A typical implementation is illustrated in Figure 7.10. A conventional transmitter without de-emphasis (Figure 7.10a) may be straightforwardly modified to provide de-emphasis, as shown in Figure 7.10b, by partitioning the output driver into segments (12 in this case) and driving some segments with delayed versions of the transmitted data. Much of the transmitter's power budget is consumed by the output drivers, which are typically required to provide a signal swing of several hundred millivolts to 50 Ω loads, in which case the overhead of the digital multiplexors in Figure 7.10b is negligible. The circuitry required to generate the delayed data signals $d(k-1)$ and $d(k-2)$ is not shown in Figure 7.10, but at worst comprises one or more latches clocked at the data rate.

It is clear that the peak transmitted voltage in Figure 7.10b remains the same regardless of the multiplexor settings (which define the FIR filter response) since the maximum current that can be delivered to the load always remains equal to the sum of all 12 output driver segments. Hence, this implementation of transmit equalization automatically maintains a constant peak transmitted voltage while providing a flexible and digitally programmable frequency response. The overhead associated with this implementation, compared to a transmitter with no equalization, comprises mainly the digital multiplexors whose power consumption should be much smaller than that of the output stages anyway.

A disadvantage of transmit equalization, if the filter parameters are to be made adaptive, is that a mechanism must be provided for communicating information about the received waveform back to the transmitter. There is also another, more subtle disadvantage in terms of performance in the performance of the link in the presence of noise, which is discussed in the next subsection.

7.4.2 Receive Equalization

Forward equalization at the receiver may be performed by either a discrete-time filter [6–8] or a continuous-time filter [19, 20, 22] preceding the decision device. Unlike a preemphasis filter, which processes digital inputs, a linear receive equalizer must be capable of handling the full dynamic range of signals that can appear at the receiver front end, generally making it a more challenging circuit design.

For the link dominated by independent noise in Figure 7.9 the signal at the output of the receiver front end comprises the desired signal from the far-end transmitter, $\upsilon(t)$, plus a noise component, $w(t)$, where

$$\upsilon(t) = (u * b * h * c)(t) \tag{7.12}$$

and

$$\omega(t) = (x * c)(t) \tag{7.13}$$

Assume that for a given channel the transmitter and receiver must combine to provide a desired frequency response, $K \cdot E(\omega)$, which equalizes the channel. Therefore,

$$B(\omega) \cdot C(\omega) = K \cdot E(\omega) \tag{7.14}$$

Without loss of generality, the frequency-dependent term $E(\omega)$ satisfies $|E(\omega)| \leq 1$ and the gain constant K normalizes the received signal amplitude. The transmitter and receiver impulse responses must therefore satisfy

$$(b * c)(t) = K \cdot e(t) \tag{7.15}$$

where $e(t)$ is the inverse Fourier transform of $E(\omega)$.

Combining (7.12) and (7.15), it becomes clear that using either transmit de-emphasis, a receive-side linear equalizer, or any combination of the two satisfying (7.14) will yield the exact same noiseless received signal,

$$\upsilon(t) = K \cdot (u * e * h)(t) \tag{7.16}$$

This is not surprising since the channel and equalizer are ideally linear systems, so the order in which they appear has no effect on the end-to-end pulse response. The noise, however, only passes through the receiver, and so has a power spectral density given by

$$S_w(\omega) = S_x(\omega) \cdot |C(\omega)|^2 \tag{7.17}$$

where $S_a(w)$ is the power spectral density of $a(t)$. Therefore, to minimize the noise power, it is necessary to minimize the magnitude response of the receiver (thereby attenuating the noise) while still satisfying (7.14) to equalize the channel. If the transmitter is peak-constrained to an amplitude of L, it must satisfy (7.11). So, subject to (7.11) and (7.14), the noise power at the receiver (7.17) is minimized as follows:

$$|B(\omega)| = L \tag{7.18}$$

$$|C(\omega)| = \frac{K}{L}|E(\omega)| \tag{7.19}$$

So, when the transmitter is peak-constrained, SNR at the receiver is maximized by using a transmitter with a flat-frequency response (hence, no de-emphasis) and performing all of the equalization at the receiver. In many high-speed serial links, the dominant noise is actually the superposition of many independent noise sources and is well approximated by a bounded Gaussian distribution [14]. In this case, maximizing the SNR also minimizes the bit error rate.

In summary, assuming ideal implementations of transmit and receive equalization are possible and putting aside their complexity and power consumption, receive equalization offers better performance than peak-constrained transmit equalization (de-emphasis). Nevertheless, transmit equalization is often preferred for its simpler implementation. Furthermore, the performance advantages of receive equalization can become very small in parallel links, as discussed in the next section.

7.4.3 CROSSTALK

In a parallel bus of chip-to-chip links, the primary source of noise is often self-crosstalk from neighboring channels. Since, as shown in Figure 7.11, the primary noise source is influenced by the transmitter design, very little performance advantage is offered by receive equalization.

Specifically, let $u_2(t)$ equal the signal being communicated over the link while $u_1(t)$ is an independent aggressor. If, as before, the combination of transmitter and receiver is required to have a frequency response of $K \cdot E(\omega)$ (7.14), then any combination of transmit- or receive-side equalization will yield the same noiseless received signal,

$$v(t) = (u_2 * b * h * c)(t)$$
$$= K \cdot (u_2 * e * h)(t) \tag{7.20}$$

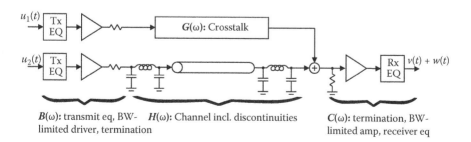

$B(\omega)$: transmit eq, BW- $H(\omega)$: Channel incl. discontinuities $C(\omega)$: termination, BW-
limited driver, termination   limited amp, receiver eq

FIGURE 7.11 System models for a parallel chip-to-chip link limited by self-crosstalk equalization.

However, unlike independent noise links, the aggressors in self-crosstalk links are communicating data using a similar transmitter, $B(\omega)$. Therefore, regardless of where the equalization is performed, the noise at the output of the receiver front end will be exactly the same:

$$\omega(t) = (u_1 * b * g * c)(t)$$
$$= K \cdot (u_1 * e * g)(t) \tag{7.21}$$

Equations (7.20) and (7.21) show that the performance of a link limited by self-crosstalk is the same using transmit de-emphasis, receiver equalization, or any combination of the two. This is intuitive since the transmitter and receiver appear in both the signal path and the dominant noise path. The result is also applicable in the presence of transmitter and receiver jitter since these also affect both the signal and noise paths.

In practice, parallel links are subject to both independent and self-crosstalk noise. Hence, receive equalization still offers some performance advantage, albeit very small when the noise is dominated by self-crosstalk.

7.5 MODULATION

Simple binary non-return-to-zero (NRZ) signaling is still commonly used for most chip-to-chip interfaces, primarily due to its straightforward hardware implementation. However, baseband pulse amplitude modulation (PAM) with four [3, 23], six [24], and eight levels [11] has been suggested. A common question is, when should multilevel modulation be used instead of binary signaling?

As in all band-limited channels, bandwidth-efficient signaling schemes are required to extend the data rates over chip-to-chip interconnect significantly beyond 2× the channel bandwidth. In this context, what *precisely* is the channel bandwidth? This question will be considered for serial and parallel links separately.

7.5.1 SERIAL LINKS

Using 4-PAM halves the operating frequency compared with binary signaling, resulting in reduced channel losses. However, the level spacing is also reduced, by one-third or 9.5 dB. So, 4-PAM is thought to be advantageous whenever the reduced level spacing is compensated for by improved signal transmission and lower noise.

A common approach is to quantify the improvement in signal transmission and the reduction in noise obtained by going to 4-PAM using the ratio of the noise spectrum to the channel response (noise-to-channel ratio, *NCR*):

$$NCR(\omega) = \frac{S_x(\omega)}{|H(\omega)|^2} \tag{7.22}$$

If one is comparing binary and 4-PAM for communication at a data rate of f_b, then the simple rule states that 4-PAM is beneficial whenever

$$20 \log_{10} \left(\frac{NCR(\pi f_b)}{NCR(\pi f_b/2)} \right) > 9.5 \text{ dB} \tag{7.23}$$

Assuming NCR decreases linearly with increasing frequency on a log-log plot, Equation (7.23) indicates that 4-PAM signaling is preferable to binary signaling as long as the slope of the NCR curve exceeds 9.5 dB per octave (31.5 dB per decade) at one-half the data rate [4, 8]. The argument is straightforwardly extended to conclude that M-PAM is preferable to N-PAM whenever

$$20 \log_{10} \left(\frac{NCR(2\pi f_b/\log_2(N))}{NCR(2\pi f_b/\log_2(M))} \right) > 20 \log_{10} \left(\frac{(M-1)}{(N-1)} \right) \text{dB} \tag{7.24}$$

Unfortunately, the preceding analysis is somewhat oversimplified, as demonstrated by simulations in, for example, [14] and [25]. Its greatest shortcoming is that it fails to recognize the broadband nature of baseband communication signals. "Spot" measures of NCR at $f_b/2$ and $f_b/4$ are insufficient to characterize the performance of binary and 4-PAM links. Since baseband binary signals have considerable spectral content at low frequencies, they can be very reliable even when the channel response is quite poor at one-half the bit rate. Hence, the straightforward criteria described above unfairly favor 4-PAM signals.

Advocates of 4-PAM signaling also often correctly point out that less high-frequency boost is required to equalize a 4-PAM signal than a binary signal at the same data rate since the binary spectrum extends to higher frequencies where channel losses increase. However, due to their reduced level spacing, 4-PAM signals are more sensitive to residual ISI. Hence, in practice, 4-PAM signals may actually be more difficult to equalize since greater accuracy is required in defining the equalizer response.

7.5.2 PARALLEL BUSES

In parallel bus chip-to-chip links dominated by self-crosstalk, deciding which modulation to use is complicated by the fact that both the signal and noise are influenced by the transmitted spectrum.

This problem has been heavily researched for communication over digital subscriber loops where crosstalk from multiple subscribers is a dominant source of noise [26]. Under these circumstances, it has been shown that the transmitted spectrum that maximizes the total capacity of all links is given by the solution to a quadratic equation:

$$|G(\omega)|^2 \left(|G(\omega)|^2 + |H(\omega)|^2 \right) S_u^2(\omega) + S_x(\omega) \left(2|G(\omega)|^2 \right.$$
$$\left. + |H(\omega)|^2 \right) S_u(\omega) + S_x(\omega) \left(S_x(\omega) - A + |H(\omega)|^2 \right) = 0 \tag{7.25}$$

In (7.25), the noise power includes both crosstalk from channels with the same transmitted spectrum, $S_u(\omega)|G(\omega)|^2$, and other noise, $S_x(\omega)$.

The results presented in [27] extend this analysis to consider peak-power-limited PAM communication, which is more applicable to chip-to-chip links than the Gaussian distributions considered in [26]. In both cases, the analysis indicates that the transmit spectrum should be mostly restricted to frequencies where $|H(\omega)|^2/|G(\omega)|^2 \gtrsim 1$. The procedure, then, is to choose a modulation method so that the Nyquist rate is below the frequency at which $|H(\omega)|^2/|G(\omega)|^2 = 1$.

A particularly interesting result of this analysis is that multilevel modulation is generally unnecessary for purely far-end crosstalk (FEXT)-limited channels. Because the crosstalk is also attenuated by the channel's through response, it is impossible to have $|H(\omega)|^2/|G(\omega)|^2 \gtrsim 1$ at any frequency.

7.6 CONCLUSION

Modern integrated chip-to-chip communication transceivers include various combinations of transmitter de-emphasis, receiver equalization, and multilevel modulation. These alternatives were analyzed and their performance compared.

Transmit- and receive-side linear equalization were found to offer equivalent performance when crosstalk from synchronized links with similar transmitters is the dominant outside interference. However, receive-side linear equalization is preferable when other (independent) noise sources are dominant. These conclusions are also applicable to the linear portion of a decision feedback equalizer. Ideal circuit implementations were assumed to compare the achievable performance using both approaches. In practice, receive-side equalizers are a more difficult circuit design since they must process the full dynamic range of the received signal, whereas transmit equalizers accept quantized (often binary digital) inputs. However, receive-side equalizers are more easily adapted online when no reverse channel is present.

A comparison of binary and multilevel modulation was also presented. The common practice of using the slope of a channel's *NCR* curve to predict the relative performance of binary and 4-PAM signaling was shown to be somewhat oversimplified. It unfairly favors multilevel modulation. Furthermore, multilevel modulation was found to be generally undesirable for purely FEXT-limited channels. The problem of finding the optimal transmit spectrum for a given chip-to-chip link remains an open question requiring further study.

Of course, as the data rates over a given channel are increased, eventually multilevel signaling becomes necessary. Consider the migration of voice-band modems to more bandwidth-efficient modulation schemes throughout the 1980s and 1990s. However, bandwidth-efficient modulation demands more sophisticated equalization than binary signaling. For example, a 4-PAM transmitter in [28] employed 13 taps of de-emphasis to compensate for 14.5 dB loss at one-half the symbol rate. The cost associated with the additional silicon area and power consumption of such sophisticated equalization could, instead, be applied to lowering the channel loss by employing a low dielectric [29].

APPENDIX

Transmission matrices provide a description of linear electrical two-port networks that are particularly useful for modeling complex chip-to-chip links. In this appendix, mathematical transformations are provided from several other electrical two-port network descriptions to transmission matrices.

IMPEDANCE MATRICES

Referring to Figure 7.4, impedance matrices relate the terminal voltages and currents of a two-port as follows:

$$\begin{bmatrix} V_1 \\ V_2 \end{bmatrix} = \begin{bmatrix} Z_{11} & Z_{12} \\ Z_{21} & Z_{22} \end{bmatrix} \begin{bmatrix} I_1 \\ I_2 \end{bmatrix} \tag{7.26}$$

A transmission matrix describing the same two-port is given by

$$\Phi = \frac{1}{Z_{21}} \begin{bmatrix} Z_{11} & \Delta Z \\ 1 & Z_{22} \end{bmatrix} \tag{7.27}$$

where

$$\Delta Z = Z_{11} Z_{22} - Z_{12} Z_{21} \tag{7.28}$$

ADMITTANCE MATRICES

Referring to Figure 7.4, admittance matrices relate the terminal voltages and currents of a two-port as follows:

$$\begin{bmatrix} I_1 \\ I_2 \end{bmatrix} = \begin{bmatrix} Y_{11} & Y_{12} \\ Y_{21} & Y_{22} \end{bmatrix} \begin{bmatrix} V_1 \\ V_2 \end{bmatrix} \tag{7.29}$$

A transmission matrix describing the same two-port is given by

$$\Phi = -\frac{1}{Y_{21}} \begin{bmatrix} Y_{22} & 1 \\ \Delta Y & Y_{11} \end{bmatrix} \tag{7.30}$$

where

$$\Delta Y = Y_{11} Y_{22} - Y_{12} Y_{21} \tag{7.31}$$

SCATTERING PARAMETERS

Scattering parameters are a linear two-port network description often obtained from laboratory measurements on a network analyzer. The four frequency-dependent complex-valued scattering parameters S_{11}, S_{12}, S_{21}, and S_{22} are related to the entries of an equivalent transmission matrix as follows:

$$A = \frac{(1+S_{11})(1-S_{22})+S_{12}S_{21}}{2S_{21}} \tag{7.32}$$

$$B = Z_0 \cdot \frac{(1+S_{11})(1+S_{22})-S_{12}S_{21}}{2S_{21}} \tag{7.33}$$

$$C = \frac{1}{Z_0} \cdot \frac{(1-S_{11})(1-S_{22})-S_{12}S_{21}}{2S_{21}} \tag{7.34}$$

$$D = \frac{(1-S_{11})(1+S_{22})+S_{12}S_{21}}{2S_{21}} \tag{7.35}$$

RLGC VALUES

Per-unit-length values of series resistance R, series inductance L, shunt conductance G, and shunt capacitance C are sometimes derivable from a knowledge of the physical geometry and material properties of a transmission line. Once available, these may be transformed into the transmission matrix of a transmission line segment of length d,

$$\Phi = \begin{bmatrix} \cosh(gd) & Z_0 \cdot \sinh(gd) \\ \dfrac{1}{Z_0} \cdot \sinh(gd) & \cosh(gd) \end{bmatrix} \tag{7.36}$$

where the characteristic impedance is

$$Z_0 = \sqrt{\frac{(R+jwL)}{(G+jwC)}} \tag{7.37}$$

and the propagation constant is

$$g = \sqrt{(R+jwL)(G+jwC)} \tag{7.38}$$

ACKNOWLEDGMENTS

Sincerest thanks and appreciation to Mike Bichan for providing the measurement results of backplane channels and Amer Samarah for his help with the figures and appendix.

REFERENCES

1. D. V. Plant and A. G. Kirk, Optical interconnects at the chip and board level: challenges and solutions, *Proc. IEEE*, 88(6), 806–818, 2000.
2. M.-C. F. Chang, V. P. Roychowdhury, L. Zhang, H. Shin, and Y. Qian, RF/wireless interconnect for inter- and intra-chip communications, *Proc. IEEE*, 89(4), 456–466, 2001.
3. R. Farjad-Rad, C. K. Yang, M. Horowitz, and T. H. Lee, A 0.3-μm CMOS 8-Gb/s 4-PAM serial link transceiver, *IEEE J. Solid-State Circuits*, 35(5), 757–764, 2000.
4. J. Zerbe, C. Werner, V. Stojanovic, F. Chen, J. Wei, G. Tsang, D. Kim, W. Stonecypher, A. Ho, T. Thrush, R. Kollipara, M. Horowitz, and K. Donnelly, Equalization and clock recovery for a 2.5–10 Gb/s 2-PAM/4-PAM backplane transceiver cell, *IEEE J. Solid-State Circuits*, 38(12), 2121–2130, 2003.
5. A. Martin, B. Casper, J. Kennedy, J. Jaussi, and R. Mooney, 8 Gb/s differential simultaneous bidirectional link with 4 mv 9 ps waveform capture diagnostic capability, in *IEEE International Solid-State Circuits Conference*, February 2003, pp. 78–79.
6. J.-Y. Sim, Y.-S. Sohn, H.-J. Park, C.-H. Kim, and S.-I. Cho, 840 Mb/s CMOS demultiplexed equalizing transceiver for DRAM-to-processor communication, in *Symposium on VLSI Circuits*, June 1999, pp. 23–24.
7. J. E. Jaussi, G. Balamurugan, D. R. Johnson, B. K. Casper, A. Martin, J. T. Kennedy, N. Shanbhag, and R. Mooney, An 8 Gb/s source-synchronous I/O link with adaptive receiver equalization, offset cancellation and clock deskew, *IEEE J. Solid-State Circuits*, 40(1), 80–88, 2005.
8. Y. Hur, M. Maeng, C. Chun, F. Bien, H. Kim, S. Chandramouli, E. Gebara, and J. Laskar, Equalization and near-end crosstalk (NEXT) noise cancellation for 20-Gb/s 4-PAM backplane serial I/O interconnections, *IEEE. Trans. Microwave Theory Tech.* 53(1), 246–255, 2005.
9. Y.-S. Sohn, S.-J. Bae, H.-J. Park, and S.-I. Cho, A 2.2 Gbps CMOS look-ahead DFE receiver for multidrop channel with pin-to-pin time skew compensation, in *Custom Integrated Circuits Conference*, September 2003, pp. 473–476.
10. R. Payne, B. Bhakta, S. Ramaswamy, S. Wu, J. Powers, P. Landman, U. Erdogan, A.-L. Yee, R. Gu, L. Wu, Y. Xie, B. Parthasarathy, K. Brouse, W. Mohammed, K. Heragu, V. Gupta, L. Dyson, and W. Lee, A 6.25Gb/s binary adaptive DFE with first post-cursor tap cancellation for serial backplane communications, in *IEEE International Solid-State Circuits Conference*, February 2005, pp. 68–69.
11. D. J. Foley and M. P. Flynn, A low-power 8-PAM serial transceiver in 0.5-μm digital CMOS, *IEEE J. Solid-State Circuits*, 37(3), 310–316, 2003.
12. J. L. Zerbe, P. S. Chau, C. W. Werner, W. F. Stonecypher, H. J. Liaw, G. J. Yeh, T. P. Thrush, S. C. Best, and K. S. Donnelly, A 2 Gb/s/pin 4-PAM parallel bus interface with transmit crosstalk cancellation, equalization, and integrating receivers, in *IEEE International Solid-State Circuits Conference*, February 2001, pp. 66–67.
13. K. Chang, S. Pamarti, K. Kaviani, E. Alon, S. Shi, T. J. Chin, J. Shen, G. Yip, C. Madden, R. Schmitt, C. Yuan, F. Assaderaghi, and M. Horowitz, Clocking and circuit design for a parallel I/O on a first-generation CELL processor, in *IEEE International Solid-State Circuits Conference*, February 2005, pp. 526–527.
14. V. Stojanovic and M. Horowitz, Modeling and analysis of high-speed links, in *Custom Integrated Circuits Conference*, September 2003, pp. 589–594.
15. E. Guizzo and H. Goldstein, Expressway to your skull, *IEEE Spectrum*, 43(8), 34–39, 2006.
16. J. E. Jaussi, B. K. Casper, M. Mansuri, F. O'Mahony, K. Canagasaby, J. Kennedy, and R. Mooney, A 20Gb/s embedded clock transceiver in 90nm CMOS, in *IEEE International Solid-State Circuits Conference*, February 2006, pp. 340–341.
17. A. X. Widmer and P. A. Franaszek, A dc-balanced, partitioned-block 8B/10B transmission code, *IBM J. Res. Dev.*, 27(5), 440–451, 1983.

18. H. Johnson and M. Graham, *High-speed signal propagation: advanced black magic.* Englewood Cliffs, NJ: Prentice Hall, 2003.
19. N. Krishnapura, Q. Barazande-Pour, M. Chaudhry, J. Khoury, K. Lakshmikumar, and A. Aggarwal, A 5Gb/s NRZ transceiver with adaptive equalization for backplane transmission, in *IEEE International Solid-State Circuits Conference*, February 2005, pp. 60–61.
20. M. Sorna, T. Beukema, K. Selander, S. Zier, B. Ji, P. Murfet, J. Mason, W. Rhee, H. Ainspan, and B. Parker, A 6.4Gb/s CMOS SerDes core with feedforward and decision-feedback equalization, in *IEEE International Solid-State Circuits Conference*, February 2005, pp. 62–63.
21. K. Krishna, D. A. Yokoyama-Martin, S. Wolfer, C. Jones, M. Loikkanen, J. Parker, R. Segelken, J. L. Sonntag, S. Titus, and D. Weinlader, A 0.6 to 9.6Gb/s binary backplane transceiver core in 0.13μm CMOS, in *IEEE International Solid-State Circuits Conference*, February 2005, pp. 64–65.
22. S. Gondi, J. Lee, D. Takeuchi, and B. Razavi, A 10Gb/s CMOS adaptive equalizer for backplane applications, in *IEEE International Solid-State Circuits Conference*, February 2005, pp. 328–329.
23. J. T. Stonick, G.-Y. Wei, J. L. Sonntag, and D. K. Weinlader, An adaptive PAM-4 5 Gb/s backplane transceiver in 0.25 μm CMOS, *IEEE J. Solid-State Circuits*, 38(3), 436–443, 2003.
24. K. Farzan and D. A. Johns, A low-complexity power-efficient signaling scheme for chip-to-chip communication, in *IEEE International Symposium on Circuits and Systems*, May 2003, vol. 5, pp. 77–80.
25. B. K. Casper, M. Haycock, and R. Mooney, An accurate and efficient analysis method for Gb/s chip-to-chip signaling schemes, in *Symposium on VLSI Circuits,* June 2002, pp. 54–57.
26. I. Kalet and S. Shamai, On the capacity of a twisted-wire pair: Gaussian model, *IEEE Trans. Commun.*, 38(3), 379–383, 1990.
27. S. Shamai, On the capacity of a twisted-wire pair: peak-power constraint, *IEEE Trans. Commun.*, 38(3), 368–378, 1990.
28. A. E. A. Amirkhany, A 24Gb/s software programmable multi-channel transmitter, in *Symposium on VLSI Circuits*, June 2007, pp. 38–39.
29. B. Casper, G. Balamurugan, J. E. Jaussi, J. Kennedy, M. Mansuri, F. O'Mahony, and R. Mooney, Future microprocessor interfaces: analysis, design and optimization, presented at Custom Integrated Circuits Conference, September 2007.

8 25 G/40 G SerDes: Need, Architecture, and Implementation

Rohit Mittal

CONTENTS

A SerDes or serializer/deserializer is an integrated circuit (IC or chip) transceiver that converts parallel data to serial data and vice versa. The transmitter section is a serial-to-parallel converter, and the receiver section is a parallel-to-serial converter. Multiple SerDes interfaces are often housed in a single package. For the remainder of this chapter, SerDes and physical layer (PHY) will be used interchangeably.

The basic SerDes function is made up of two functional blocks: the parallel in, serial out (PISO) block (aka transmitter) and the serial in, parallel out (SIPO) block (aka receiver), as shown in Figure 8.1.

The PISO (parallel input, serial output) block typically has a clock input and a set of data input lines synchronous to the clock input. It may use an internal or external phase-locked loop (PLL) to multiply the incoming clock up to the serial frequency. The simplest form of the PISO has a single shift register that receives the parallel data once per clock, and shifts it out at the higher serial clock rate.

FIGURE 8.1 SerDes block diagram.

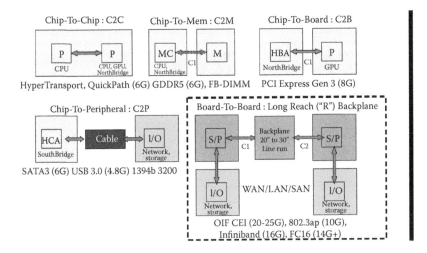

FIGURE 8.2 SerDes applications.

The SIPO (serial input, parallel output) block typically has a receiver that takes in the input data, recovers a high-frequency synchronous clock, and spits out parallel data at a lower-frequency clock rate. The receive clock may have been recovered together with the data by the clock recovery techniques explained later in the chapter. The simplest form of SIPO has two registers connected as a double buffer. One register is used to clock in the serial stream, and the other is used to hold the data for the slower, parallel side.

SerDes spans many different applications, as shown in Figure 8.2.

The focus of this chapter will be the boxed application since this is the place where the highest speed will occur first (as shown later).

8.1 25 AND 40 G SERDES: NEEDS AND TRENDS

Increasingly higher-bandwidth requirements continue to drive development and demand for 40 G and 100 G systems. Examples of consumer applications include YouTube, Facebook, smartphones, and IP-TV. Governmental and business demands compound the urgency of higher bandwidth with a variety of complex data-intensive solutions, including weather prediction, financial analysis, genomics research, and design simulation. Further, the rapid emergence of cloud computing for both personal and business use adds additional demands for high-volume, high-complexity data transmission.

Figures 8.3 and 8.4 shows the huge growth in networking requirements. It also shows that most of the data are processed within the data center—between switches or switches and servers. It is here that 25 G/40 G links will be needed.

Why is there a need for higher serial data instead of just more links of parallel data? Higher speed reduces the cost of data transmission since there are fewer pipes that need to be built for the same amount of information transfer. Examples of this include the migration of Sonet from Oc12 to Oc768.

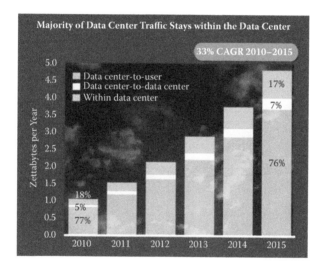

FIGURE 8.3 Growth of data center traffic. (From Cisco.)

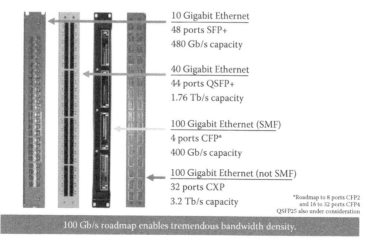

FIGURE 8.4 Effect of I/O bandwidth on traffic in data centers. (From LSI.)

We can see that there has been a gradual upward trend in the top data rate of electrical transceivers, as shown in Figure 8.5.

The highest reported speed for a complementary metal-oxide-semiconductor (CMOS) electrical transceiver, 40 Gb/s, occurred in 2009 [2] and again in 2011 [3]. While higher speeds have been reported in exotic III-V material such as SiGe, CMOS shows enough promise to become the mainstream technology for 25 G/40 G.

So where will 25 G and 40 G SerDes be used? The key driver, as mentioned before, will be the data center market. The sweet spot is 100 Gb/s transmission. Initially 100 Gb/s was implemented with 10 × 10 G due to the maturity of 10 G CMOS IC and modules. Now this is progressing to 25 G/40 G, with a few recent papers highlighting 25 G/40 G in CMOS.

Data are sent from the application-specific integrated circuit (ASIC) on the line card via a PHY on a backplane. The backplane has connectors that connect to a module. The module will contain a retimer block (4:4 PHY) that feeds into an optical module as shown in Figure 8.6. This will ultimately go on a single-mode (SMF) or multimode (MMF) fiber to another module in a switch rack, for example.

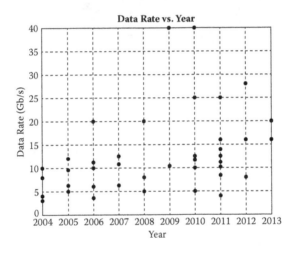

FIGURE 8.5 Published data rate per year. (From J. Stonick, ISSCC 60th anniversary supplement, 2013.)

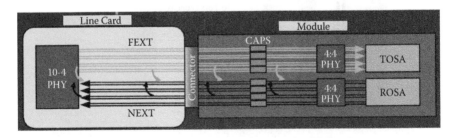

FIGURE 8.6 System level view of high-speed links in data networking equipment. (From Inphi.)

8.2 MEETING THE NEEDS VIA 25 G/40 G STANDARDS

For 25 G/40 G the two main standards bodies are OIF and Ethernet [4–6]. There are others such as Fibre Channel, but the main difference is in the protocol layer rather than the SerDes layer. Furthermore, OIF and Ethernet are quite similar in terms of output and input specifications (electrical specifications). Since mixed-signal circuits are primarily used in the SerDes layer, only OIF and Ethernet specifications will be considered going forward.

The OIF standard is only a physical layer standard, and the portion relevant to this chapter will be the electrical interface to the optical modules. The optical modules by themselves will not be discussed since that technology is based on lasers and photodetectors, which are discussed elsewhere in the book.

CEI-25 G LR = 25 G over long reach.
CEI-28 G SR = 28 G over short reach.
CEI-28 G VSR = 28 G over very short reach.

The Institute of Electrical and Electronics Engineers (IEEE) standard includes both the logical layer, physical medium attachment (PMA) and the physical layer, physical medium dependent (PMD). This chapter will primarily focus on the PMA-PMD interface and the electrical PMD—the high-speed SerDes layers. This is the first time two different speeds (100 G/40 G) have been specified in a single IEEE 802.3 standard [4]. The decision to include both speeds comes from pressure to support the 40 Gbit/s rate for local server applications and the 100 Gbit/s rate for Internet backbones. The standard was announced in July 2007 and was ratified on June 17, 2010.

Every SerDes needs to be compliant to one of the standards mentioned in Table 8.1. System requirements for SerDes have become increasingly demanding due to link speeds and new standards, yet 40 G/100 G transmissions must retain backwards compatibility. In addition, the SerDes needs to consume the least amount of power in the minimum amount of space.

Shown in Tables 8.2 and 8.3 is part of the OIF-28 G SR specifications [6].

Translating Table 8.2 into numbers, duty cycle distortion (DCD) needs to be less than 1.25 ps, and the total jitter requirement is 10 ps. Due to bit error rate (BER) requirements of 1e-15, random jitter gets multiplied by ±8 sigma to provide the

TABLE 8.1
Present Ethernet Standards for 40 G/100 G

Physical Layer	40 Gigabit Ethernet	100 Gigabit Ethernet
Backplane	40 GBASE-KR4	
Copper cable	40 GBASE-CR4	100 GBASE-CR10
100 m over OM3 MMF	40 GBASE-SR4	100 GBASE-SR10
125 m over OM4 MMF		
10 km over SMF	40 GBASE-LR4	100 GBASE-LR4
40 km over SMF		100 GBASE-ER4
Serial SMF over 2 km	40 GBASE-FR	

TABLE 8.2
Transmitter Electrical Output Specification

Characteristic	Symbol	Condition	MIN.	TYP.	MAX.	UNIT
Baud rate	T_Baud		19.90		28.05	Gsym/s
Output differential voltage	T_Vdiff	Emphasis off. See Note 4	800		1200	mVppd
Differential resistance	T_Rd		80	100	120	Ω
Differential termination resistance mismatch (see Tables 8.1 and 8.2)	T_Rdm				10	%
Output rise and fall time (20% to 80%)	T_tr, T_tf	Emphasis off. See Note 2	8			ps
Common mode noise	T_Ncm	Note 3			12	mVrms
Differential output return loss	T_SDD22	See Section 10.3.1.3				dB
Common mode output return loss	T_SCC22	Below 10 GHz 10 GHz to baud rate			−6 −4	dB
Output common mode voltage	T_vcm	Load type 0 See Note 1	−100		1700	mV
Uncorrelated unbounded gaussian jitter	T_UUGJ				0.15	UI_{pp}
Uncorrelated bounded high probability jitter	T_UBHPJ	Note 2			0.15	UI_{pp}
Duty cycle distortion (component of UBHPJ)	T_DCD	Note 3			0.035	UI_{pp}
Total jitter	T_TJ	Note 1			0.28	UI_{pp}

Notes: 1. T_TJ includes all of the jitter components measured without any transmit equalization.
2. Measured with all possible values of transmitter equalization, excluding DDJ as defined in 12.1.1.
3. Included in T_UBHPJ.

TABLE 8.3
Receiver Electrical Input Specifications

Characteristic	Symbol	Condition	MIN.	TYP.	MAX.	UNIT
Baud rate	R_Baud		19.90		28.05	GSym/s
Input differential voltage	R_Vdiff	Note 1			1200	mVppd
Differential input impedance	R_Rdin		80	100	120	Ω
Input impedance mismatch	R_Rm				10	%
Differential input return loss	R_SDD11	See 10.3.2.3				
Common mode input return loss	R_SCC11	Below 10 GHz 10GHz to baud rate			−6 −4	dB
Input common mode voltage	R_Vcm	Load Type 0 See Note 2	−200		1800	mV

UIpp number listed in the table. Hence, random jitter (RJ) has to be ~330 fs (rms), a nontrivial task.

Similar specifications exist for the CEI-25 G-LR standard.

In addition to meeting the strict 100 GBASE-LR4/ER4 jitter specification, the 25 Gb/s SerDes transceivers used in 100 Gb/s line cards will have to overcome the signal degradation caused by the impedance mismatches, nonlinear attenuation, crosstalk, and other channel impairments commonly encountered on the printed circuit boards used in networking equipment. To counteract the effects of intersymbol interference, modern transceivers use equalization techniques at both the transmitter and the receiver of the SerDes link. Building such equalizers is nontrivial, especially with power and area constraints.

As mentioned before, the published state of the art architectures are able to meet these specifications in 28 to 40 nm CMOS through innovative design techniques while retaining the architecture derived in this chapter. The next section will look into the implementation details of ultra-high speed CMOS SerDes. Note to the reader: Apart from mixed-signal circuits shown in the next few sections, there are other technologies needed to make long range 100 G systems feasible such as FEC (forward error correction), PAM4 (pulse amplitude modulation) etc. These are beyond the scope of this chapter.

8.3 MIXED-SIGNAL SERDES IMPLEMENTATION FOR 25 G/40 G

Fundamentally, a SerDes can be broken into three components:

1. Transmitter (TX)
2. Receiver (Rx)
3. Clocking

A simple picture is shown in Figure 8.7.

The 25 G/40 G SerDes has some enhancements over the fundamental block diagram and is shown in Figure 8.8. Each of these enhancements will be discussed

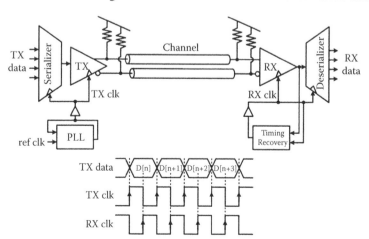

FIGURE 8.7 Typical SerDes components.

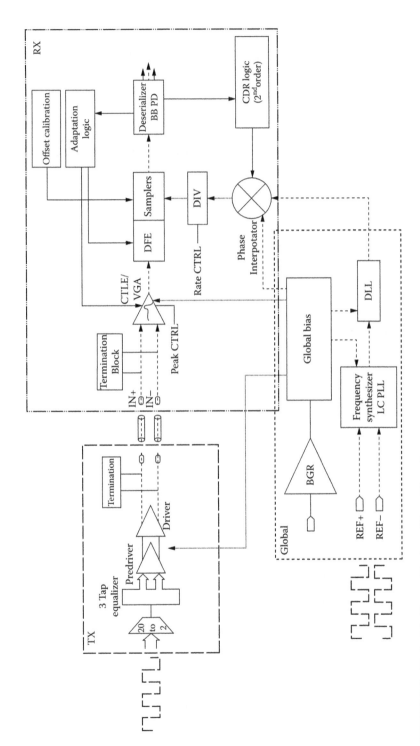

FIGURE 8.8 Typical SerDes architecture. (From Mosys.)

in detail in the subsequent sections. Global blocks are typically instantiated only once in a SerDes implementation. On the other hand, there may be many instances of Rx and Tx—one per lane in a SerDes. So if the SerDes is four lanes (to provide an aggregate of 100 G), there will be 4 Tx and 4 Rx but only one global block.

Given the large demand and exciting challenges of these ultra-high-speed systems, there is a lot of research going on in this area. Implementations shown in the subsequent sections are primarily based on typical work as well as extrapolation of the present state of the art. There are implementations that are not covered (due to space constraints), as well as widely different from the ones covered below.

8.4 GLOBAL

Global lanes contain a bandgap reference that will generate the current and voltages that are sent to other parts of the SerDes (i.e., data lanes). A sample low-voltage, low-power bandgap reference can be used here [10]. The global lane also contains the central LCPLL—an inductor-based PLL.

8.4.1 LCPLL

As mentioned before, the jitter need for 25 G standards is <300 fs rms jitter. An ultra-low jitter LCPLL is needed to meet the exacting requirements.

LCPLL is a critical piece of the SerDes since it creates the majority of SerDes random jitter (RJ). All the other blocks typically contribute deterministic jitter (DJ) but minimal RJ. Furthermore, high-frequency RJ cannot be tracked by the receiver.

The block diagram is shown in Figure 8.9. $G(s)$ is the open-loop response of a stable Gardner style PLL with two poles (at DC), one zero and one pole above the unity gain frequency of the PLL.

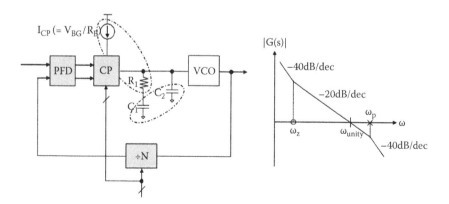

FIGURE 8.9 PLL block diagram.

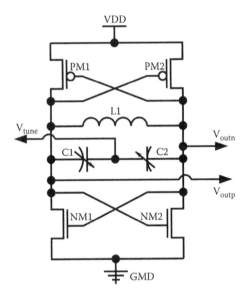

FIGURE 8.10   A typical LC VCO (fully complementary).

The critical part of the jitter comes from the VCO, which is shown in Figure 8.10.

There has been fundamental work done on the nature of LC VCO noise and its derivation as a linear time-varying (LTV) system [7]. While the math of LTV phase noise is quite involved, a design-friendly equation for the VCO phase noise is as follows:

$$L(f_m) = \frac{1}{8Q^2} \frac{FkT}{C} \left(\frac{f_o}{f_m} \right)^2$$

where F is the circuit noise factor, k is Boltzman's constant, T is temperature, C is the oscillator output power, Q is the tank quality factor, f_o is the oscillator fundamental frequency, and f_m is the frequency offset from the fundamental.

At 25 G+ speeds, the keys to low jitter are the use of a supply regulator with better than 20 dB of supply rejection and the inherent spectral purity of the LC oscillator via a high-Q inductor design.

The $L(f_m)$ is converted to RJ of the PLL via phase noise integration [8].

8.5   TRANSMITTER (Tx)

Creating a 25 G/40 G Tx has the following challenges:

1. Low rise and fall time/high bandwidth. This is necessary for reducing intersymbol interference (ISI)-induced jitter.
2. Impedance control. This is needed for reducing reflection-induced jitter.
3. Low capacitive load. This is to reduce return loss as well as ISI.
4. Low power.

Tx can be built using current mode logic (CML) or voltage mode (VM) architectures. A high-speed Tx has typically been built with CML architecture. A CML architecture, shown in Figure 8.11, allows lower-voltage operation relative to push-pull, which leads to lower power.

Recently voltage mode drivers have started to show up in very high-speed SerDes as well (Figure 8.12).

Theoretically, voltage mode drivers have one-fourth the power of current mode drivers (assuming similar predriver power). Unfortunately, the low matching resistance requirements cause a large area requirement for the drivers. This nullifies some of the power savings from going to VM techniques. Nevertheless, VM drivers have good potential and this was highlighted as a major source of innovation [9].

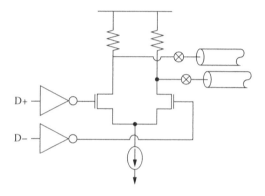

FIGURE 8.11   A typical CML Tx.

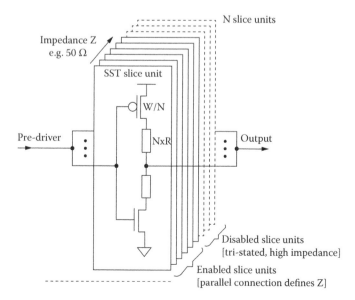

FIGURE 8.12   A voltage mode Tx.

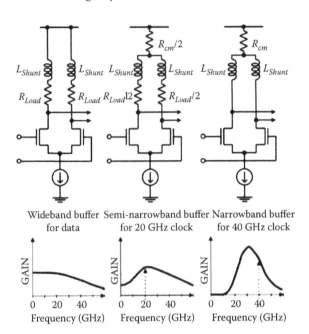

FIGURE 8.13 Methods to increase speed of operation in Tx for 25 G/40 G. (From S. Kaeriyama et al., *IEEE Journal of Solid-State Circuits*, 44(12), 3568–3579, 2009.)

As speed increases toward the ultra-high end (40 Gb/s), inductive tuning is used extensively for compensating the load capacitance effects on bandwidth. As shown in Figure 8.13, both series and shunt inductors have been used in [11].

Apart from the final Tx stage, the previous stages to generate the 25 G/40 G data stream are also challenging. These multiplexing stages provide the PISO function and also implement the finite impulse response (FIR) filter digitally. They are typically implemented in a custom digital environment due to the high speeds involved (the earlier stages can be realized via auto place and route). The high-speed design requires careful layout planning as well as intensive circuit simulation over process, voltage, temperature (PVT) corners to ensure robust design (Figure 8.14) [12].

8.6   RECEIVER (Rx)

Creating a 25 G/40 G Rx is very challenging. Apart from the challenges of the Tx mentioned before, the Rx also needs equalizers to mitigate the effects of the channel. While there has been work on analog-to-digital converter (ADC)-based equalization for SerDes before [13], the present state of the art 25 G+ SerDes primarily uses analog equalization techniques.

The equalizers for such high speed consist of not only a feed-forward continuous-time linear equalizer (CTLE) but also a nonlinear decision feedback equalizer (DFE). To compound the complexity further, these equalizers are typically adaptive to compensate for the varying environmental conditions on the high-speed system. Frequently, the Rx is the most complicated block in the SerDes.

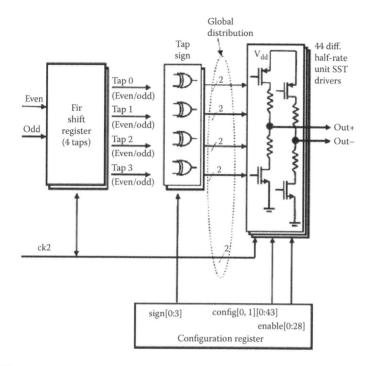

FIGURE 8.14 Implementation of high-speed FIR filter in the digital domain. (From T. Toifl et al., Low-Power High-Speed CMOS I/Os: Design Challenges and Solutions, presented at TWEPP, Ruschlikon, Switzerland, September 2012.)

8.6.1 RECEIVER EQUALIZERS

Attenuation through the link/channel causes intersymbol interference (ISI). The ISI is mitigated via equalizers. The equalizer can be in the Tx or Rx section, as shown in Figure 8.15.

The complexity of the equalizer depends upon the channel characteristics that need to be compensated. A short-distance route may need a simple or no equalizer. In addition, the equalizer may be best situated in the Tx where the implementation is easier. Certainly this is historically the place where most SerDes equalization techniques start.

However, longer-distance channels and higher-speed channels (25 G+) with more dispersive characteristics need complicated equalizers. A Tx-only equalizer may not suffice since this limits the input swing to the receiver. Hence, an equalizer will be used in the receiver as well. If the dispersion is very large, a DFE may be needed as well.

As an example, the channel from the line card to the module corresponds to OIF-25 G-LR or 802.3ba over backplane. This will require equalization. The channel from 4:4 PHY is a shorter distance and similar to the OIF-CEI-Min specification, as shown in Figure 8.16. This application will not require equalization or may have a very simple one.

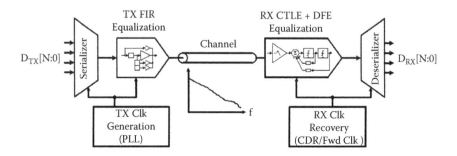

FIGURE 8.15   Equalization and channel attenuation.

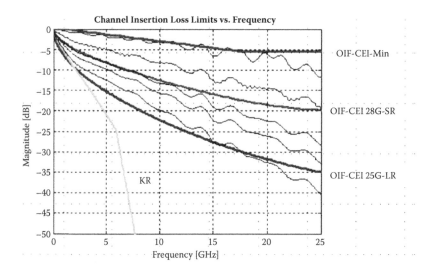

FIGURE 8.16   CTLE mitigation of typical channel attenuation.

For ultra-high speeds, a simple CTLE runs into issues because the parasitic pole (w_{p2}) will end up in the desired passband. While it is feasible to burn extra power by reducing the load resistor to push out the parasitic pole, there is a limit to its efficacy, since larger devices would be needed to handle this increased current.

Some known methods prototyped to get around this issue include using discrete-time equalizers [3] and inductive peaking [14]. The latter suffers from a large area increase (as is common in any on-chip inductor scheme).

While a CTLE does an admirable job in narrowing the receiver pulse, it also amplifies the input noise. This is due to the high-pass nature of the circuit. The only way to equalize without noise amplification is to use a DFE, as shown in Figure 8.17.

A typical circuit implementation of DFE uses current steering logic to implement the taps, h_{-1} in Figure 8.17. The delay stages, depicted as T in the figure, are implemented as flip-flops. At high speed, typical circuit issues similar to CTLE come into play—parasitic caps, self-loading from switches. Some of these issues are resolved

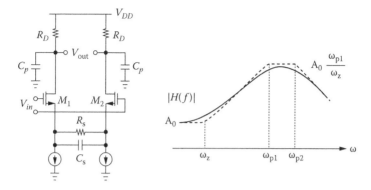

FIGURE 8.17 Conceptual view of DFE.

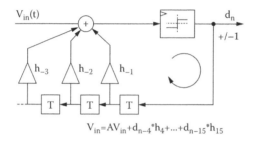

$$V_{in} = AV_{in} + d_{n-4}^*h_4 + \ldots + d_{n-15}^*h_{15}$$

FIGURE 8.18 Closing the first tap feedback in DFE. (From S. Kaeriyama et al., *IEEE Journal of Solid-State Circuits*, 44(12), 3568–3579, 2009.)

by using switched capacitors to replace the resistors, as shown in Figure 8.18 [15]. However, there are also fundamental architecture issues associated with DFE.

For high speed and ultra-high speed, closing the timing around the first tap is nontrivial. For higher speed, loop unrolling is used along with soft decision making, as shown in Figure 8.19 [11].

Unfortunately, there is exponential growth in complexity as the number of unrolled taps increases [16]. Building DFE for 25 G+ links is one of the most challenging tasks in a SerDes [17].

8.6.2 Receiver Samplers

If the equalizers do their job well, the output will be open enough that it can be resolved by samplers into a digital high or low. At 25 G and long channels, the eye will be closed at the entry point of the receiver. The equalizers will ensure there is a small enough window for every unit interval (UI) to distinguish between 1 and 0. The key role for the sampler is to amplify the equalized waveform into a digital output that can then be de-multiplexed by the subsequent Rx stages.

Invariably, the samplers have been implemented as strong-arm flops [18]. These flops use regeneration/positive feedback to attain higher speeds, as shown in Figure 8.20.

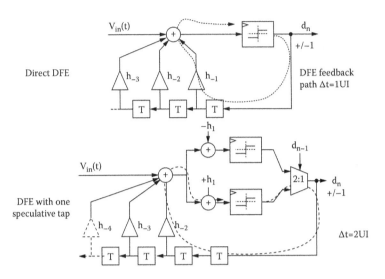

Direct DFE

DFE feedback
path Δt=1UI

DFE with one
speculative tap

Δt=2UI

FIGURE 8.19 Loop unrolling in DFE [11].

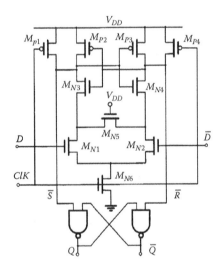

FIGURE 8.20 Regenerative strong-arm flip flop.

As the data rate increases to the ultra-high end, the signal coming into the sampler becomes smaller and smaller (due to larger channel attenuation as well as difficulty in getting gain from previous equalizer stages). This requires careful control of sampler offsets.

The receiver also contains many other blocks, such as clock data recovery (CDR), phase interpolators (PIs), adaptation and calibration engines, signal/idle detection blocks, and analog biasing circuits. In particular, the first two blocks, CDR and PI, are very challenging to implement due to the varying needs from the standards *vs.* stability requirements [19].

FIGURE 8.21　Implementation of multilane SerDes [21].

8.7　PUTTING IT ALL TOGETHER

Once all the blocks are designed as per some of the techniques outlined in the previous section, they need to be assembled together for the final IP/product. There is a wealth of material regarding the challenges and techniques to build a robust mixed-signal flow for designing and validating the SerDes, in both the front end [20] and the back end. Figure 8.21 shows an example of the final physical implementation of the SerDes IP. Other configurations may be possible where the chip logic is above the lanes rather than on the side (for a larger number of data lanes).

8.8　CONCLUSION

The chapter began by looking at the exploding demand of higher speeds and the need of 25 G/40 G SerDes to meet this demand. This was followed by an overview of the standards in this space and the architectures to meet those standards.

We then looked at the various blocks in the SerDes and the challenges in building the blocks. Finally, we built the blocks and put them together into a complete physical implementation of a high-speed SerDes.

ACKNOWLEDGMENTS

This tutorial chapter would not have been feasible without the support of multiple reviewers. In particular, the author thanks Lluis Paris, Navraj Nandra, Mian Quddus, Fulvio Spagna, Hiok-Tiaq Ng, Sanjay Sethi, and Gaurav Malhotra.

REFERENCES

1. J. Stonick, ISSCC 60th anniversary supplement, 2013.
2. Y. Amamiya et al., A 40Gb/s multi-data-rate CMOS transceiver chipset with SFI- 5 interface for optical transmission systems, *ISSCC Digest of Technical Papers*, February 2009, pp. 358–359, 359a.

3. M.-S. Chen et al., A 40Gb/s TX and RX chipset in 65nm CMOS, *ISSCC Digest of Technical Papers*, February 2011, pp. 146–148.
4. http://en.wikipedia.org/wiki/IEEE_802.3ba.
5. http://en.wikipedia.org/wiki/100_Gigabit_Ethernet.
6. http://www.oiforum.com/public/documents/OIF_CEI_03.0.pdf.
7. T. H. Lee and A. Hajimiri, Oscillator phase noise: a tutorial, *IEEE Journal of Solid-State Circuits*, 35(3), 326–336, 2000.
8. R. Mittal, Tutorial: clock jitter measurement and effects, *EE Times*, June 20, 2011, pp. 41–42.
9. C. Menolfi et al., A 16Gb/s source-series terminated transmitter in 65nm CMOS SOI, *ISSCC Digest of Technical Papers*, February 2007, pp. 446–614.
10. B Hironori et al., A CMOS bandgap reference circuit with sub-1-V operation, *IEEE Journal of Solid-State Circuits*, 34, 670–674, 1999.
11. S. Kaeriyama et al., A 40 Gb/s multi-data-rate CMOS transmitter and receiver chipset with SFI-5 interface for optical transmission systems, *IEEE Journal of Solid-State Circuits*, 44(12), 3568–3579, 2009.
12. T. Toifl et al., Low-power high-speed CMOS I/Os: design challenges and solutions, presented at TWEPP, Ruschlikon, Switzerland, September 2012.
13. M. Harwood et al., A 12.5Gb/s SerDes in 65nm CMOS using a baud-rate ADC with digital receiver equalization and clock recovery, *ISSCC Digest of Technical Papers*, February 2007, pp. 436–591.
14. D. Li et al., A 25Gb/s low noise 65nm CMOS receiver tailored to 100GBASE-LR4, in *ESSCIRC*, 2012, pp. 221–224.
15. M. Park et al., A 7 Gb/s 9.3 mW 2-tap current integrating DFE receiver, presented at ISSCC 2007.
16. T. Toifl et al., A 3.1mW/Gbps 30Gbps quarter-rate triple-speculation 15-tap SC-DFE RX data path in 32nm CMOS, in *IEEE Symposium on VLSI Circuits*, June 2012, pp. 102–103.
17. V. Telang, Equalization for high speed serdes, http://ewh.ieee.org/r5/denver/sscs/Presentations/2012_08_Telang.pdf. 2012.
18. B. Nikolic et al., Improved sense-amplifier-based flip-flop: design and measurements, *IEEE Journal of Solid-State Circuits*, 35(6), 876–884, 2000.
19. J. Stonick, DPLL-based clock and data recovery, ISSCC tutorial, February 2011.
20. Designers' consulting website, http://www.designers-guide.com/.
21. F. Zhong et al., A 1.0625 14.025 Gb/s multi-media transceiver with full-rate source-series-terminated transmit driver and floating-tap decision-feedback equalizer in 40 nm CMOS, *IEEE Journal of Solid-State Circuits*, 46(12), 3126–3138, 2011.

9 Clock and Data Recovery Circuits

Jafar Savoj

CONTENTS

9.1   INTRODUCTION

The raw speed and low cost of modern complementary metal-oxide-semiconductor (CMOS) technologies has fueled the evolution of integrated circuits into large systems-on-chip (SOC). The increase in the aggregate clocking speed and the number of supported features in an SOC require high-speed data transceivers at the input/output (I/O) ports, operating at speeds as high as a few tens of gigabits per second. Traditional implementations of high-speed transceivers utilized various techniques to achieve maximum operating rate and highest signal purity despite the speed limitations of available fabrication processes. These techniques included using external components, additional mask layers through the fabrication process, and choice of high-power logic families. Other implementations benefited from exotic fabrication technologies that provided higher raw speeds but did not achieve low cost and low power dissipation. Low yield of devices fabricated in such processes did not allow the integration of the entire system and made the circuits only appropriate for use in multichip systems.

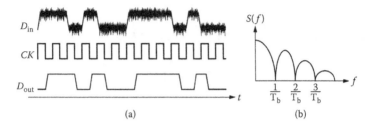

FIGURE 9.1   (a) Role of the CDR circuit, (b) spectrum of random data.

Modern designs benefit from much higher process speeds and target full integration of transceivers in very large scale integration (VLSI) technologies. Such implementations consider critical issues including small area and power consumption of the core, as well as provisions to allow the use of low-cost package and printed circuit board. On-chip signal processing techniques are devised to reverse the dispersive effects of these materials stemming from poor impedance control of the lines. The design of the clock and data recovery (CDR) circuit is the most challenging task in implementation of any wireline receiver. To fulfill stringent requirements for power consumption, integrability, and ease of portability across process nodes, new architectures have transitioned from fully analog into hybrid and digital structures. In this chapter, we study the evolution of CDR architectures and describe some of their key implementations.

Shown in Figure 9.1(a), the task of the CDR circuit is to regenerate the data arriving at the receiver. The received signal experiences attenuation and distortion from the communication channel and is further corrupted by deterministic and random sources of disturbance, such as crosstalk, reflection, and supply noise. The regeneration circuit inside the receiver is synchronized with the incoming data and captures the data at the instant of maximum eye opening, typically at the center of the bit period. The sampling clock signal should achieve high purity (in terms of timing jitter) because the purity of the regenerated signal can never exceed that of the sampling clock signal. Therefore, clock recovery focuses on synchronizing the sampling clock with the data, positioning the sampling edge of the clock at the center of the data eye, and reducing the timing jitter of the clock signal to the smallest possible.

As shown in Figure 9.1(b), the power spectral density function of random data lacks frequency components at the data rate and its integer multiples. Therefore, the synchronizing clock signal cannot be directly derived from the data. Clock recovery techniques provide solutions to extract the synchronization information from the data.

In this chapter, we provide an overview of common CDR architectures and describe the implementation of their key building blocks. Next, we focus on the design of high-performance LC oscillators for CDR circuits. Finally, we study the means of characterizing jitter for CDR circuits.

9.2   INJECTION-LOCKED CDR CIRCUIT

Edge detection of random data transforms their spectrum to contain a frequency component at the frequency of the data rate. Edge detection can be implemented in the form of either differentiation and rectification, or multiplication of the signal by

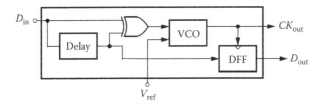

FIGURE 9.2   Injection-locked CDR circuit.

a delayed version of itself. Band-pass filtering of the resulting signal can extract the synchronizing frequency [1]. However, the center frequency-to-bandwidth ratio of such filters is very high and beyond the practical limits for integration of filters in VLSI technologies.

The major architectures suitable for implementing integrated CDR circuits were originally deployed for synchronization in television receivers. These receivers relied on injection locking of the received synchronization signal into a scanning oscillator to produce a scanning signal at the receiver [2]. Figure 9.2 depicts the utilization of injection locking in CDR circuits based on this concept. The output of the edge detector, formed with a delay element and a multiplier, is injected into a voltage-controlled oscillator (VCO). The VCO output locks to the burst of pulses arriving at its input and produces an output clock signal with the same frequency as the input. Due to its injection locking nature, this circuit acquires lock, instantaneously upon the arrival of pulses at the input, and loses lock when the pulses disappear. As a result, this architecture is only suitable for burst-mode communications [3]. The recovered clock is used to retime the data inside a D-flip-flop (DFF). This structure can only lock to the incoming pulses if the frequency of the free-running VCO is close to that of the incoming pulses. A reference voltage (V_{ref}) generated by an auxiliary loop is used to drive the VCO toward its locking range. We will later describe means of producing the reference voltage in the context of aided frequency acquisition.

9.3   PHASE-LOCKED CDR CIRCUIT

With relative poor immunity of injection-locked circuits to noise, an automatic frequency-and phase-controlled synchronizing system that relied on phase locking of a local scanning oscillator to the horizontal and vertical pulses in a television receiver was introduced in [2]. This structure resulted in improvements in television service despite severe noise conditions and expanded the coverage area. This mechanism for phase synchronization still remains a common approach for clock recovery in wireline communication systems. The original discoveries for high immunity of these circuits to noise are applicable to CDR circuits and play an important role in the jitter performance of wireline transceivers.

Figure 9.3 shows the architecture of a phase-locked CDR circuit. A phase detector (PD) produces an output that is proportional to the phase difference between the input data and the VCO (we will later see that this information can be as simple as a

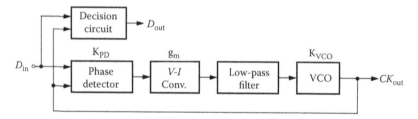

FIGURE 9.3   Phase-locked CDR circuit.

logic value to indicate whether the clock sampling edge is early or late). The combination of a voltage-to-current (*V-I*) converter and a low-pass filter (LPF), formed at a
minimum by a resistor and a capacitor in series, provides a control voltage required
to sustain the VCO oscillation at the desired phase and frequency. Finally, a decision circuit is used to retime the data using the recovered clock signal. A thorough
study into the dynamics of phase-locked systems can be found in [4]. CDR circuits
are mostly followed by a demultiplexer that converts the serial recovered data into
parallel lines.

In order to achieve minimum bit error rate (BER), either the input data or the
recovered clock has to be delayed so that the sampling edge of the recovered clock
falls in the middle of the data eye—a condition that is hard to guarantee by design
despite process and temperature variations. To achieve the optimum sampling
instant, some PD topologies not only provide the phase error information, but also
automatically retime the data. We refer to this concept as automatic retiming and
only focus on this type of PDs in this chapter.

9.3.1   FULL-RATE CDR

In a full-rate CDR circuit, the frequency of the recovered clock equals the incoming
data rate, allowing a clock rising (or falling) edge to appear in the middle of every bit.
Single-edge retiming avoids degradation of the quality of detection arising from the
duty cycle mismatch of the clock signal. We categorize the full-rate circuits based on
the type of PD used inside them.

9.3.1.1   Linear PD

In [5], Hogge developed a circuit consisting of a cascade of two flip-flops and two
XOR gates. The circuit produces two pulses at every transition of input data sequence
(Figure 9.4). One of the pulses has a width equal to the phase difference between the
data edge and the clock edge (*error*), and the other pulse has a constant width equal
to one-half of a bit period (*reference*). The difference of these two pulses is integrated
by the *V-I* converter (or a charge pump for lower-speed implementations) and the
LPF. Characterization of the integrated value versus phase error over one bit period
indicates a linear relationship. The slope of the output is an indication of the PD gain
(K_{PD}). Both inputs to the reference XOR and one input to the error XOR experience

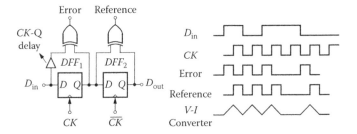

FIGURE 9.4   Hogge PD.

a CK-Q delay inside the flip-flops from the instant that the clock signal is asserted. The input data arriving at the first XOR gate are delayed to account for this effect.

When the integral of the difference falls to zero, the loop locks, indicating that the phase difference between the clock and the data has reached one-half of a bit period, and the clock samples the data in the middle of the eye. Interestingly, D_{out} is the retimed version of the input in lock condition.

Hogge PD was commonly used in CDR implementations until the late 1990s due to its simplicity and its predictable gain and delay characteristic over process and temperature. With the advancement of digital CDR solutions, binary PDs have enjoyed more popularity in recent years. However, there have been efforts to utilize the linear characteristic of Hogge PD in hybrid analog/digital solutions [6].

9.3.1.2   Binary PD

As the operation speed of CDR circuits increases, binary PDs that rely on samples of the data to provide the phase error information in the form of early or late signals, and achieve higher immunity to nonidealities of the analog process, have become more popular. Alexander PD [7] is perhaps the most commonly used binary PD. As shown in Figure 9.5(a), the operation of the circuit is based on obtaining three consecutive samples of the data, one at the edge and two from the previous and the current bits. Similarity of the edge sample to the current bit is an indication of a late clock signal (Figure 9.5(b)), and its similarity to the previous bit is an indication of early clock (Figure 9.5(c)). Three identical samples indicate that no data transition has occurred and the PD output is tristated. This feature prohibits the PD from producing a continuous early or late signal when experiencing no transitions due to a long sequence of identical bits. Comparison of data samples is performed using two XOR gates. The edge sample is obtained with the falling edge of the clock and is retimed with the rising edge. This retiming aligns B with A and C. Similar to the Hogge PD, the difference of early and late signals is integrated by the V-I converter and the LPF. The second data sample (A) is usually used as the retimed data because the cascade of two flip-flops further improves the quality of the retimed data.

In lock condition, the recovered clock dithers around data transition, producing early and late pulses such that their average effect over a time interval on the order of the inverse of the loop bandwidth is zero. In practice, metastability of flip-flops transforms the early/late characteristic of the PD into a linear characteristic with high gain.

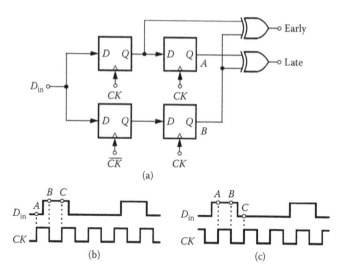

FIGURE 9.5 (a) Alexander PD, (b) late clock, and (c) early clock.

The gain of the Alexander PD is further reduced due to the jitter of the sampling clock. In [8] the actual characteristic of the PD is predicted by convolving the probability density function (PDF) of the clock jitter and the characteristic derived from the metastability analysis of the flip-flops.

9.3.2 Subrate CDR

A CDR circuit is typically followed by a demultiplexer that converts the serial data stream into parallel lines at a lower rate. Data demultiplexing requires low-frequency clock signals, produced by dividing the frequency of the recovered clock. If clock recovery could be performed using clock signals at a subrate of the input data, power consumption could be reduced due to lower-speed clocking and removal of high-power frequency dividers from the circuit. Furthermore, such schemes enable operation at higher speeds for any CMOS process node.

9.3.2.1 Linear PD

Linear PDs can be implemented to operate with subrate clock signals to relax speed constraints. A half-rate linear architecture only requires differential phases of the clock, a unique feature distinguishing this structure from a binary scheme requiring quadrature phases to operate. Shown in Figure 9.6, the circuit consists of four latches and two XOR gates [9]. Data are applied to the inputs of two sets of cascaded latches, each constituting a flip-flop that retimes the data. Since the flip-flops are driven by a half-rate clock, the two output sequences are the demultiplexed waveforms of the input sequence if the clock samples the data in the middle of the bit period.

The output of each latch tracks its input for half a clock period and holds the value for the other half, resulting in the waveforms shown in Figure 9.6 for X_1 and X_2. The XOR of these two pulses consists of pulses that appear when a data transition occurs.

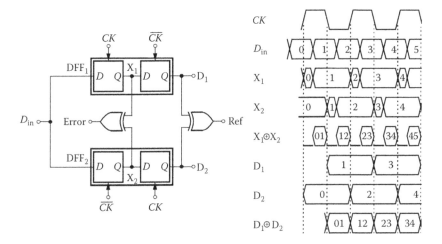

FIGURE 9.6 Half-rate linear PD.

The width of these pulses is determined by the clock edge on one side and data transition on the other, indicating that the average value of the pulse sequence contains information about the phase difference. However, the random nature of the data and the periodic behavior of the clock make this average value pattern dependent. Therefore, a reference signal must also be generated whose average conveys this dependence.

The two waveforms D_1 and D_2 contain the samples of the data at the rising and falling edges of the clock. Thus, the XOR of D_1 and D_2 contains pulses as wide as half the clock period for every data transition, serving as the reference signal. The height of the error signal should be scaled by a factor of two with respect to the reference, since the error is half as wide as the reference in the lock condition.

9.3.2.2 Binary PD

Shown in Figure 9.7(a), an Alexander PD obtains the edge and bit samples using the rising and falling edges of a full-rate clock. If half-rate clock phases are utilized, the same samples can be obtained using the in-phase and quadrature clock signals. Similarly, a quarter-rate PD relies on half-quadrature clock phases to obtain similar samples. Since mismatch between multiple phases of the clock degrades the performance of subrate circuits, utilization of these circuits is usually limited to half-rate and quarter-rate operation.

Figure 9.7(b) depicts a half-rate implementation of the Alexander PD [10]. The two data samples are obtained using the rising and falling edges of the in-phase clock. The quadrature clock signal obtains the edge sample. Similar to the full-rate approach, two XOR gates compare these samples to derive early and late information. A modification to this circuit includes introduction of an extra flip-flop, retiming C with CK_I, in order to provide alignment of the edge sample with data samples to improve the performance of the circuit.

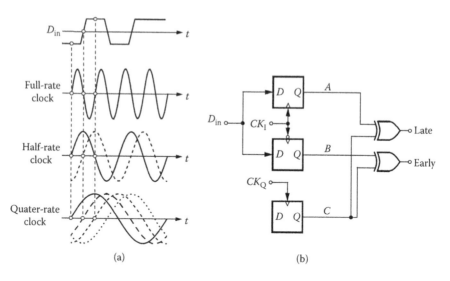

(a)   (b)

FIGURE 9.7   (a) Subrate sampling, (b) half-rate binary PD.

9.3.3 AIDED FREQUENCY ACQUISITION

The oscillator used inside the CDR circuit is designed to have a wide tuning range to provide reliable oscillation at the frequency of interest despite process and temperature variations. CDR loops are usually designed for a small closed-loop bandwidth to allow higher suppression of input jitter. The CDR circuit has a capture range on the order of its loop bandwidth. As a result, if the initial VCO frequency, at the moment of start-up, is outside the CDR capture range, the circuit may never acquire phase lock. Also, any sudden disturbance of the clock phase might force the loop out of lock, even if it has initially acquired lock. Therefore, phase-locked CDR circuits must utilize frequency acquisition schemes for stable operation. Acquisition circuits usually rely on a reference signal, used for clock synthesis inside the transmitter. Other schemes depend on samples of data themselves for frequency acquisition. These topologies are usually referred to as reference-less frequency acquisition circuits. Study of such circuits is beyond the scope of this chapter. Here, we consider circuits that utilize a reference signal for frequency acquisition. The dual-loop architecture shown in Figure 9.8 [11] relies on the wide capture range of a phase-frequency detector (PFD) operating with periodic signals [12] and consists of two loops sharing the charge pump (V-I converter), the LPF, and the VCO. A multiplexer selects the mode of operation by passing the output from either the PFD of loop I or the PD of loop II. When the circuit starts, loop I acquires lock to the reference signal, producing a VCO output frequency of Nf_{ref}. The difference between this frequency and the data rate is usually very small and within the acquisition range of the CDR circuit. Consequently, when loop II is enabled, the CDR circuit can acquire lock to the incoming data. The lock detector, designed in the form of a digital counter, determines the instant of transition.

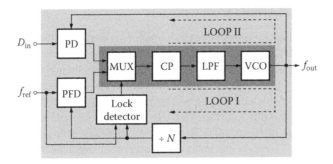

FIGURE 9.8 Dual-loop lock acquisition.

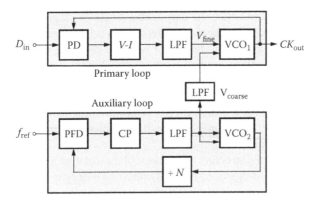

FIGURE 9.9 Lock acquisition and K_{VCO} reduction.

This topology provides acquisition with very small overhead in hardware and area because the building blocks can be shared between both loops. In a real implementation, the loop parameters should be chosen based on the requirements for loop II with attention to the stability of loop I.

Along with aided frequency acquisition, dual-loop architectures can provide flexibilities in the design of the CDR circuit. For example, the circuit proposed in [13] can be used to reduce VCO gain (K_{VCO}) in a CDR circuit and lower its timing jitter, since a smaller VCO gain translates the noise on the control line of the VCO into smaller jitter.

As shown in Figure 9.9, the structure consists of two loops utilizing identical VCOs. The control voltage of the VCO in the primary loop is split into coarse and fine lines. In this implementation, the coarse voltage is provided by the auxiliary loop. When the auxiliary loop acquires lock, it produces a control voltage required to drive the main loop into its capture range. This voltage is heavily filtered to avoid excess jitter in the primary VCO. The auxiliary loop can theoretically be turned off to save power when phase lock to random data has been achieved. However, this circuit should power up instantaneously in case of a sudden loss of lock. A similar auxiliary loop can be implemented in an injection-locked CDR topology to drive the VCO within the capture range of the circuit.

9.4   PHASE-ROTATING CDR CIRCUIT

Traditionally, high-speed transceivers have utilized separate VCOs for the transmitter and the receiver due to a slight frequency difference between the received and transmitted data. This difference stems from frequency mismatch of the far-end and local crystal oscillators. The increase in the aggregate data rate processed by an SOC can be accommodated using parallel streams of high-speed data. For example, 10 Gb/s Attachment Unit Interface (XAUI) and Quad OC-192 standards achieve total operating rates of 10 and 40 Gb/s using four transceivers in parallel, each operating at 2.5 and 10 Gb/s, respectively. The large area overhead of LC oscillators, required for high-frequency signal generation and lower jitter, is more pronounced if each transceiver has separate VCOs for the transmitter and the receiver. Therefore, means of sharing oscillators among all transceivers must be sought. Figure 9.10 demonstrates a configuration where the output of a single clock generator is routed to four transmitters. Since all four transmitters are synchronized, a shared clocking structure reduces the area. Introduction of resonant clocking schemes has significantly reduced the overall power consumption of such clocking networks [14].

The variation of incoming phase (and possibly frequency) among the received signals can only be resolved using separate CDR circuits for each receiver. To save power and area, we must consider means of utilizing the clock, produced by the clock generator as the base signal and rotating its phase to track the edges of the incoming data sequences.

Figure 9.11 depicts the structure of a generic phase-rotating architecture. Multiple phases of a clock signal produced by the synthesizer enter the phase rotator whose output clock phase is steered by the signals produced by the controller. In order to rotate the output phase along all four quadrants (full 360°), both polarities of the in-phase and quadrature clock signal (CK_I and CK_Q) must be utilized, indicating that differential signaling must be used. Phase rotation in any quadrant can be accomplished with two orthogonal phases. The control signals for the phase rotator can be either analog or digital. Digital operation is based on rotating the output phase in fine steps such that the input data edge at any point in time is bound between clock phases selected by consecutive digital codes.

The PD and the charge pump (V-I converter) used in such structures are similar to the ones used in phase-locked circuits. In a digital approach, a binary PD is preferred

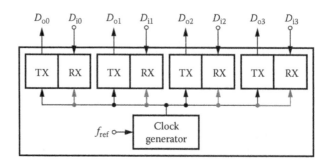

FIGURE 9.10   Clocking for multiple transceivers.

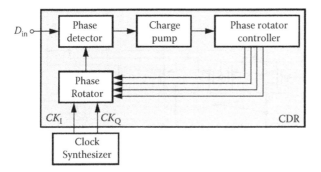

FIGURE 9.11   Phase-rotating CDR circuit.

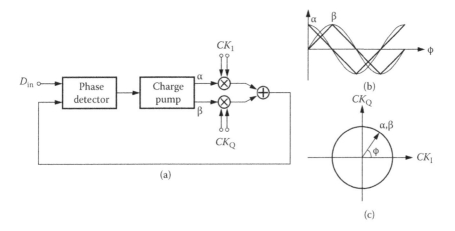

FIGURE 9.12   (a) CDR with analog phase rotator, (b) coefficients, (c) recovered clock.

and the charge pump is replaced by a digital filter. We will study digital rotation in the context of digital CDR circuits.

Figure 9.12(a) represents an analog implementation of the phase rotator [15, 16]. In this circuit, the rotator consists of two multipliers and an adder, and linearly combines the input signals in the form of $\alpha.CK_I + \beta.CK_Q$. Phase lock can be achieved if α and β represent cosine and sine of the instantaneous phase difference between the incoming data and the synthesizer frequency (Figure 9.12(b)). Note that if α and β are differential and can achieve both positive and negative values, phase rotation can be performed across all four quadrants (Figure 9.12(c)).

Shown in Figure 9.12(b), the design of the circuit can be simplified if α and β are approximated with triangular waveform. In [15], α and β signals are the differential outputs of a charge pump circuit and only possess positive values. As a result, $360°$ phase rotation across all four quadrants is accomplished by replacing a clock signal by its complement when the rotated vector passes the boundary between two quadrants. A *boundary select* circuit changes the polarity of the clock signal when its coefficient reaches zero. In [16], the circuit is modified to produce α and β signals

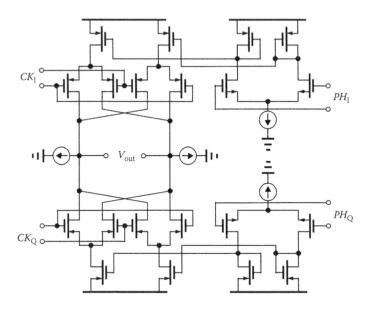

FIGURE 9.13 Analog phase rotator.

that can assume both positive and negative values, hence alleviating the requirement for clock substitution and smoothing the transition across the quadrant boundaries.

Figure 9.13 shows a low-voltage implementation of the analog phase rotator. The circuit consists of two Gilbert multipliers whose outputs are summed. In order to provide reliable low-supply operation, the phase coefficients are mirrored.

9.5   FULLY DIGITAL CDR CIRCUIT

The transition from an analog to a digital CDR architecture can be explained using the digital representation of the system, derived by substituting loop parameters in the s-domain with those in the z-domain [17]. Based on the approximation of e^{-sT} with $1 - sT$, s can be replaced in the z-domain.

$$z^{-1} = e^{-sT} \Rightarrow s \approx \frac{1 - z^{-1}}{T} \tag{9.1}$$

As a result, the combination of the analog VCO, V-I converter, and LPF can be redefined as

$$\left(\frac{K_{VCO}}{s} \right) \left(K_{PD}R + \frac{K_{PD}}{sC} \right) \rightarrow \left(\frac{K_{DPC}}{1 - z^{-1}} \right) \left(K_\phi + \frac{K_f}{1 - z^{-1}} \right) z^{-N} \tag{9.2}$$

Figure 9.14 depicts the system level representation of a digital CDR based on the transformation of Equation (9.2). In this system K_{DPC} is the gain of the digital-to-phase

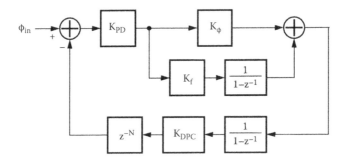

FIGURE 9.14 Digital representation of a CDR circuit.

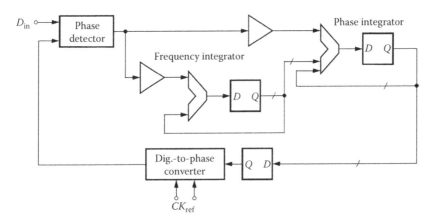

FIGURE 9.15 Digital CDR circuit.

converter (DPC), a circuit that selects one of m equally spaced clock phases according to its input digital code, and passes that to its output. The additional term given by z^{-N} represents the latency around the loop.

Figure 9.15 shows the actual implementation of the CDR circuit [17]. The PD usually has a binary subrate structure to relax speed constraints of the digital logic. The resulting early/late information is either summed or processed in a majority vote circuit, in order to reduce the width of the parallel error bus.

The frequency integrator compensates for the difference between the clock frequency and the data rate. It should have enough bits to cover the maximum frequency difference between the reference clock and the data. The gain stages are realized by shift registers. In order to reduce the volume of computation in each cycle and reduce power consumption, only the top bits of the phase and frequency integrators are passed to the proceeding stage. The discarded bits are a source of dither. The number of these bits is chosen so that the dithering of the frequency integrator is not a major source of jitter.

9.6　LC OSCILLATORS IN CDR CIRCUITS

The performance of an oscillator has a strong impact on the output jitter of the closed-loop CDR circuit. In [18] it was shown that when an oscillator is placed inside a loop with unity gain bandwidth of f_u, jitter accumulates with the square root of time and saturates thereafter. The saturated value of jitter was calculated in [19] and represented as

$$\Delta t_{CDR} = \sqrt{\frac{f_0}{4\pi f_u}} \cdot \Delta t_{VCO} \tag{9.3}$$

in which Δt_{CDR} is the total CDR jitter, f_0 is the VCO center frequency, and Δt_{VCO} is the cycle-to-cycle jitter of the open-loop VCO.

Equation (9.3) indicates that for applications with a fixed-loop bandwidth, the most effective way to reduce the output jitter is reduction of the open-loop VCO jitter. In [20] it was also shown that the phase noise due to white noise sources and the cycle-to-cycle jitter are related, indicating that VCOs should be designed with the smallest possible phase noise.

VCO phase noise, regardless of circuit topology, is a strong function of its power consumption. In general, an increase by a factor of N in power consumption is equivalent to adding the output of N identical oscillators in phase. This structure achieves an increase in the total output power by N^2, while the uncorrelated noise power only increases by N, concluding that when power is increased by N, phase noise decreases by the same factor [21]. As a result, all figures of merit defined to describe the performance of oscillators are normalized to power consumption. Despite using an area-consuming integrated inductor, LC oscillators achieve the lowest phase noise compared to other VCO topologies. Unlike ring oscillators, whose frequency is a function of inverter or differential pair delay, and highly dependent on the technology, the oscillation frequency of an LC oscillator is determined solely from the values of the integrated inductor and the total output capacitance and can be very high as long as the cross-coupled devices provide gain. LC oscillators therefore provide the highest oscillation frequency in any particular technology. Together, these features make topologies of Figure 9.16 attractive for implementation in CDR circuits.

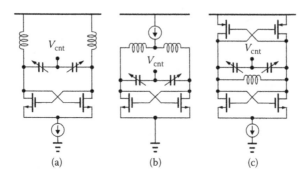

FIGURE 9.16　Conventional LC oscillators: (a) bottom current, (b) top current, (c) complementary.

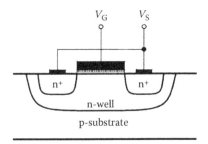

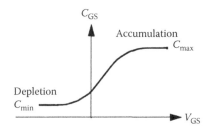

FIGURE 9.17   MOS varactors.

The oscillators of Figure 9.16 usually use an NMOS varactor to achieve frequency tunability. Formed by placing an negative-channel metal oxide semiconductor (NMOS) inside an N-well (Figure 9.17(a)), the structure achieves the maximum capacitance in accumulation and the minimum capacitance in depletion (Figure 9.17(b)). In order to achieve the maximum capacitance variability (maximum VCO tuning range) the gate-source voltage of the varactors should possess both positive and negative values.

In an LC VCO, gates of the varactors are usually connected to the output nodes and the control voltage produced by the preceding LPF is connected to its source/drain. The control voltage can typically swing close to supply and ground rails. Therefore, to achieve the maximum tuning range, the common-mode level of the VCO output should be close to mid-supply. The structure of Figure 9.16(a) can be replaced with the structures of Figure 9.16(b) and (c) to resolve this issue.

The complementary VCO structure of Figure 9.16(c) is shown to achieve the lowest phase noise for a given power consumption. This structure provides a higher g_m for a given power consumption because the current flows through both the NMOS and the positive-channel metal oxide semi-conductor (PMOS) cross-coupled pairs. Also, if the NMOS and PMOS switching devices are sized so that the VCO has identical rise and fall times, its output achieves single-ended symmetry, a phenomenon that is believed to reduce the output phase noise [22].

The study of phase noise in oscillators has been an active research topic in recent years. In [23] the major sources of phase noise in LC oscillators were identified to be the resonator noise, current source noise, differential pair noise, and amplitude modulation (AM)-to-phase modulation (PM) conversion in the varactors. Resonator noise can be reduced by designing inductors that have a higher quality factor (Q). However, it is believed that as the frequency of oscillation increases, varactors, and not the inductors, contribute to reduction of the Q of the tank. Alternative tuning techniques can be sought to alleviate the impact of low-Q varactors [24].

The current source noise translates up and down in frequency due to the switching of the cross-coupled pair and enters the resonator. As a result, components of tail current noise at DC and $2f_0$ can contribute to the output phase noise. Low-frequency noise and $2f_0$ device noise appear as amplitude noise and phase noise, respectively. Shown in Figure 9.18, an extra inductor and the capacitor can form high impedance at $2f_0$ and block the noise from entering the resonator [25]. AM-to-PM conversion represents the transformation of current source

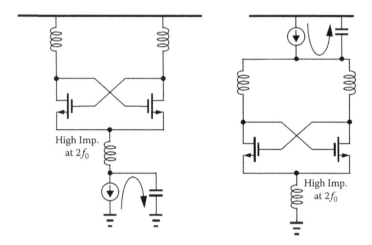

FIGURE 9.18 Noise filtering in LC oscillators.

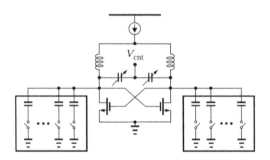

FIGURE 9.19 K_{VCO} reduction with calibration.

amplitude noise into phase noise due to voltage tunability of varactor capacitance. The varactor C-V characteristic achieves its highest slope in mid-range and its smallest slope at the two boundaries. This characteristic poses a significant design challenge because oscillators are nominally designed to operate in the center of the tuning range, where they experience the highest AM-to-PM conversion and exhibit the highest phase noise. The problem is more pronounced if the number of uncorrelated noise sources in the oscillator increases. Examples include quadrature and multiphase oscillators.

This problem can be circumvented by reducing K_{VCO}. However, this is not trivial since the VCO must have sufficient tuning range, up to a few gigahertz, running from a low-supply voltage. VCO calibration is an effective technique for K_{VCO} reduction [26]. Shown in Figure 9.19, the coarse output capacitance is selected from a bank of linear capacitors (such as fringe metal capacitors) and the fine capacitance is provided by a varactor.

When the system powers up, a calibration algorithm switches the metal capacitors and drives the VCO frequency to a range that can be covered with the fine varactors. Despite having a large tuning range, K_{VCO} of this oscillator is significantly reduced because the fine varactor constitutes a small fraction of the total output capacitance. Further studies into the operation and optimization of VCOs can be found in [27].

9.7 JITTER IN CDR CIRCUITS

Jitter in CDR circuits is characterized with three parameters. *Jitter generation* is a measure of the jitter produced by the circuit itself. The major sources of this jitter are the phase noise of the VCO, and supply and substrate noise. Utilization of LC oscillators and reduction of K_{VCO} are effective means of lowering this type of jitter.

Jitter transfer describes the closed-loop CDR circuit in terms of its unity gain bandwidth and maximum allowed peaking. The latter is a provision to avoid excessive peaking of jitter when multiple CDR circuits are cascaded in a long-haul wireline link. Characterization of transfer function is performed by modulating the edge of the input data signal with a low-frequency sinusoid and monitoring the response of the CDR circuit to this modulation. The ratio of the magnitude of the modulation frequency appearing in the recovered clock to that in the input data is an indication of jitter transfer magnitude at the frequency of modulation. Repetition of this process over various modulation frequencies produces the jitter transfer characteristic of the CDR circuit (Figure 9.20(a)) [28].

Jitter tolerance is an indication of the maximum input jitter that a CDR circuit can tolerate before it erroneously detects a bit. Figure 9.20(b) shows a jitter tolerance mask. This mask demonstrates the maximum amplitude of phase modulation that can be corrected by the loop as a function of phase modulation frequency.

Satisfaction of both jitter tolerance and transfer specifications using a single CDR loop is not trivial. Wider loop bandwidth improves trackability of the input jitter but lowers output jitter suppression. For example, a 10 Gb/s optical system requires a transfer bandwidth of 160 kHz and a tolerance bandwidth of about 4 MHz. This problem can be solved using a cleanup phase-locked loop (PLL) (Figure 9.21).

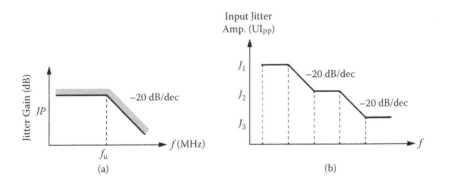

FIGURE 9.20   (a) Jitter transfer and (b) jitter tolerance masks.

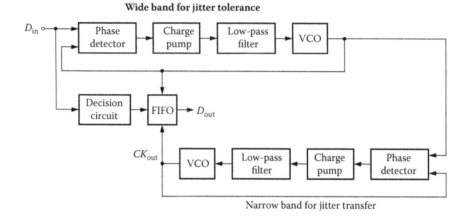

FIGURE 9.21 Cleanup PLL.

TABLE 9.1

State-of-the-Art CDR Circuits

	Speed	Process	Architecture	Power Dissipation	BER	Area
[30]	40 Gb/s	65 nm (SOI)	Quarter-rate	72 mW	$< 10^{-12}$	0.03 mm²
[31]	25 Gb/s	90 nm	Half-rate	172 mW	$< 10^{-12}$	0.07 mm²
[32]	16 Gb/s	90 nm	Subrate	35 mW	$< 10^{-10}$	0.05 mm²
[33]	10 Gb/s	90 nm	Phase-rotating	300 mW*	$< 10^{-9}$	0.43 mm²
[34]	6.25 Gb/s	0.13 μm	Digital	152 mW*	$< 10^{-15}$	0.56 mm²ᵃ
[35]	1.6 Gb/s	0.13 μm	Digital	12 mW	$< 10^{-12}$	0.1 mm²
[36]	1.5 Gb/s	90 nm	Digital	1 mW	NA	0.023 mm²
[37]	33.8 Gb/s	90 nm	Injection-locked	98 mW	NA	0.64 mm²
[38]	20 Gb/s	90 nm	Injection-locked	175 mW	$< 10^{-9}$	0.96 mm²

* Quoted for the entire transceiver.

In this circuit, the main CDR loop is designed to have a wide bandwidth for jitter tolerance. The output of this loop is the input to another PLL with a much narrower bandwidth. The cascade of these two loops achieves the narrow bandwidth, required for jitter transfer. In order to achieve the same low bandwidth for the recovered data, the circuit uses a first in, first out (FIFO), in which write and read clock signals are provided by the main CDR and the cleanup PLL, respectively. This block removes the additional jitter on the recovered data arising from retiming with the clock signal generated by the wide bandwidth loop [29].

Table 9.1 summarizes the key features of some of the recently published CDR circuits. These prototypes are considered the state of the art at the time of this publication.

9.8 CONCLUSIONS

Design of the CDR circuit is the most challenging task in the implementation of a wireline receiver. With the advancement in CMOS processes, analog architectures have transitioned into hybrid and digital solutions. New techniques have enabled sharing of a single clock generator for multiple transceivers, integrated into the same chip, and have resulted in significant reduction of power consumption and chip area.

LC oscillators, despite consuming a larger area compared to their ring counterparts, achieve much higher speeds of oscillation and superior timing jitter. Novel design methodologies have enabled their utilization in the design of high-speed CDR circuits. Also, circuit and architecture level innovations are used to fulfill stringent requirements for generation, transfer, and tolerance of jitter in modern integrated CDR circuits.

REFERENCES

1. A. W. Buchwald, Design of Integrated Fiber-Optic Receivers Using Heterojunction Bipolar Transistors, Ph.D. thesis, University of California, Los Angeles, January 1993.
2. K. R. Wendt, G. L. Fredendall, Automatic Frequency and Phase Control of Synchronization in Television Receivers, *Proceedings of the IRE*, January 1943, pp. 7–15.
3. M. Nogawa, K. Nishimura, S. Kimura, T. Yoshida, T. Kawamura, M. Togashi, K. Kumozaki, Y. Ohtomo, A 10 Gb/s Burst-Mode CDR IC in 0.13 μm CMOS, *ISSCC Digest of Technical Papers*, February 2005, pp. 228–229.
4. B. Razavi, ed., *Monolithic Phase-Locked Loops and Clock Recovery Circuits*, Piscataway, NJ: IEEE Press, 1996.
5. C. Hogge, A Self-Correcting Clock Recovery Circuit, *IEEE Journal of Lightwave Technology*, LT-3, 1312–1314, 1985.
6. M. H. Perrott, Y. Huang, R. T. Baird, B. W. Garlepp, L. Zhang, J. P. Hein, A 2.5Gb/s Multi-Rate 0.25μm CMOS CDR Utilizing a Hybrid Analog/Digital Loop Filter, *ISSCC Digest of Technical Papers*, February 2006, pp. 328–329.
7. J. D. H. Alexander, Clock Recovery from Random Binary Data, *Electronics Letters*, 11, 541–542, 1975.
8. J. Lee, K. S. Kundert, B. Razavi, Analysis and Modeling of Bang-Bang Clock and Data Recovery Circuits, *IEEE Journal of Solid-State Circuits*, 39, 1571–1580, 2004.
9. J. Savoj, B. Razavi, A 10-Gb/s CMOS Clock and Data Recovery Circuit with a Half-Rate Linear Phase Detector, *IEEE Journal of Solid-State Circuits*, 36, 761–768, 2001.
10. M. Rau, T. Oberst, R. Lares, A. Rothermel, R. Schweer, N. Menoux, Clock/Data Recovery PLL Using Half-Frequency Clock, *IEEE Journal of Solid-State Circuits*, 32, 1156–1159, 1997.
11. G. Gutierrez, S. Kong, B. Coy, 2.488 Gb/s Silicon Bipolar Clock and Data Recovery IC for SONET (OC-48), in *Proceedings of Custom Integrated Circuits Conference*, May 1998, pp. 575–578.
12. C. A. Sharpe, A 3-State Phase Detector Can Improve Your Next PLL Design, in *EDN*, September 1976, pp. 55–59.
13. J. C. Scheytt, G. Hanke, U. Langmann, A 0.155, 0.622, and 2.488 Gb/s Automatic Bit Rate Selecting Clock and Data Recovery IC for Bit Rate Transparent SDH Systems, *ISSCC Digest of Technical Papers*, February 1999, pp. 348–349.
14. R. Palmer, J. Poulton, W. J. Dally, J. Eyles, A. M. Fuller, T. Greer, M. Horowitz, M. Kellam, F. Quan, F. Zarkeshvari, A 14mW 6.25 Gb/s Transceiver in 90nm CMOS for Serial Chip-to-Chip Communications, *ISSCC Digest of Technical Papers*, February 2007, pp. 440–441.

15. T. H. Lee, K. S. Donnelly, J. T. C. Ho, J. Zerbe, M. G. Johnson, T. Ishikawa, A 2.5V CMOS Delay-Locked Loop for an 18 MBit, 500 Megabyte/s DRAM, *IEEE Journal of Solid-State Circuits*, 29, 1491–1496, 1994.

16. F. Yang, J. H. O'Neill, D. Inglis, J. Othmer, A CMOS Low-Power Multiple 2.5–3.125-Gb/s Serial Link Macrocell for High IO Bandwidth Network ICs, *IEEE Journal of Solid-State Circuits*, 37, 1813–1821, 2002.

17. J. L. Sonntag, J. Stonick, A Digital Clock and Data Recovery Architecture for Multi-Gigabit/s Binary Links, *IEEE Journal of Solid-State Circuits*, 41, 1867–1875, 2006.

18. J. A. McNeil, Jitter in Ring Oscillators, *IEEE Journal of Solid-State Circuits*, 32, 870–879, 1997.

19. B. Razavi, *Design of Integrated Circuits for Optical Communications*, New York: McGraw Hill, 2003.

20. F. Hertzel, B. Razavi, A Study of Oscillator Jitter Due to Supply and Substrate Noise, *IEEE Transactions on Circuits and Systems II: Analog and Digital Signal Processing*, 46, 56–62, 1999.

21. B. Razavi, A Study of Phase Noise in CMOS Oscillators, *IEEE Journal of Solid-State Circuits,* 31, 331–343, 1996.

22. A. Hajimiri, T. H. Lee, Design Issues in CMOS Differential LC Oscillators, *IEEE Journal of Solid-State Circuits*, 34, 717–724, 1999.

23. J. J. Rael, A. A. Abidi, Physical Processes of Phase Noise in Differential LC Oscillators, in *Proceedings of Custom Integrated Circuits Conference*, May 2000, pp. 569–572.

24. A. Maxim, A Varactor-Less 10GHz CMOS LC-VCO for Optical Communications Transceiver SOCs Using Caged Inductors, in *Proceedings of Custom Integrated Circuits Conference*, September 2006, pp. 663–670.

25. E. Hegazi, H. Sjoland, A. A. Abidi, A Filtering Technique to Lower LC Oscillator Phase Noise, *IEEE Journal of Solid-State Circuits*, 36, 1921–1930, 2001.

26. H. Darabi, S. Khorram, E. Chien, M. Pan, S. Wu, S. Moloudi, J. C. Leete, J. J. Rael, M. Syed, R. Lee, B. Ibrahim, M. Rofougaran, A. Rofougaran, A 2.4GHz CMOS Transceiver for Bluetooth, *ISSCC Digest of Technical Papers*, February 2001, pp. 200–201.

27. D. Ham, A. Hajimiri, Concepts and Methods in Optimization of Integrated LC VCOs, *IEEE Journal of Solid-State Circuits*, 36, 896–909, 2001.

28. L. M. DeVito, A Versatile Clock Recovery Architecture and Monolithic Implementation, invited paper in *Monolithic Phase-Locked Loops and Clock Recovery Circuits, Theory and Design*, ed. B. Razavi, New York: IEEE Press, 1996.

29. L. DeVito, Clock Recovery and Data Retiming, presented at MEAD Workshop on IC Design for Optical Communication Systems, Monterey, CA, March 2001.

30. T. Toifl, C. Menolfi, P. Buchmann, C. Hagleitner, M. Kossel, T. Morf, J. Weiss, M. Schmatz, A 72mW 0.03mm^2 Inductorless 40Gb/s CDR in 65nm SOI CMOS, *ISSCC Digest of Technical Papers*, February 2007, pp. 226–227.

31. C. Kromer, G. Sialm, C. Menolfi, M. Schmatz, F. Ellinger, H. Jackel, A 25 Gb/s CDR in 90nm CMOS for High-Density Interconnects, *ISSCC Digest of Technical Papers*, February 2006, pp. 326–327.

32. S. Palermo, A. E. Neyestanak, M. Horowitz, A 90nm CMOS 16Gb/s Transceiver for Optical Interconnects, *ISSCC Digest of Technical Papers*, February 2007, pp. 44–45.

33. M. Meghelli, S. Rylov, J. Bulzacchelli, W. Rhee, A Rylyakov, H. Ainspan, B. Parker, M. Beakes, A. Chung, T. Beukema, P. Pepeljugoski, L. Shan, Y. Kwark, S. Gowda, D. Friedman, A 10Gb/s 5-tap-DFE/4-Tap-FFE Transceiver in 90nm CMOS, *ISSCC Digest of Technical Papers*, February 2006, pp. 80–81.

34. K. Krishna, D. Yokoyama-Martin, S. Wolfer, C. Jones, M. Loikkanen, J. Parker, R. Segelken, J. Sonntag, J. Stonick, S. Titus, D. Weinlader, A 0.6 to 9.6Gb/s Binary Backplane Transceiver Core in 0.13mm CMOS, *ISSCC Digest of Technical Papers*, February 2005, pp. 64–65.

35. P. K. Hanumolu, M. G. Kim, G.-Y. Wei, U.-K. Moon, A 1.6 Gbps Digital Clock and Data Recovery Circuit, *Proceedings of Custom Integrated Circuits Conference*, September 2006, pp. 603–606.

36. K.-H. Chao, P.-Y. Wang, T.-H. Hsu, 0.0234mm2/1mW DCO Based Clock/Data Recovery for Gbit/s Applications, *Digest of Symposium on VLSI Circuits*, June 2007, pp. 132–133.

37. L.-C. Cho, C. Lee, S.-I. Liu, A 33.6-33.8Gb/s Burst-Mode CDR in 90nm CMOS, *ISSCC Digest of Technical Papers*, February 2007, pp. 48–49.

38. J. Lee, A 20Gb/s Burst-Mode CDR Circuit Using Injection-Locking Technique, *ISSCC Digest of Technical Papers*, February 2007, pp. 46–47.

Index